Freud's Wishful Dream Book

ALEXANDER WELSH

Freud's Wishful Dream Book

PRINCETON UNIVERSITY PRESS, PRINCETON, N.J.

Copyright © 1994 by Princeton University Press
Published by Princeton University Press, 41 William Street,
Princeton, New Jersey 08540
In the United Kingdom: Princeton University Press,
Chichester, West Sussex

Library of Congress Cataloging-in-Publication Data

Welsh, Alexander.
Freud's wishful dream book / Alexander Welsh.
p. cm.
Includes bibliographical references and index.

ISBN 0-691-03718-3 (cloth)
1. Freud, Sigmund, 1856–1939. Traumdeutung. 2. Dream
interpretation. 3. Psychoanalysis. I. Title.
BF175.5.D74W45 1994
154.6'34—dc20 94-10758

This book has been composed in Adobe Garamond

Princeton University Press books are printed on
acid-free paper and meet the guidelines for permanence
and durability of the Committee on Production
Guidelines for Book Longevity of the
Council on Library Resources

Printed in the United States of America

1 3 5 7 9 10 8 6 4 2

The whole thing is planned on the model of an imaginary walk. At the beginning, the dark forest of authors (who do not see the trees), hopelessly lost on wrong tracks. Then a concealed pass through which I lead the reader—my specimen dream with its peculiarities, details, indiscretions, bad jokes—and then suddenly the high ground and the view and the question: which way do you wish to go now?

Sigmund Freud to Wilhelm Fliess, 6 August 1899

CONTENTS

THIS SHORT BOOK consists of a commentary in five parts on Freud's *The Interpretation of Dreams,* as published in the last months of 1899 and revised over the years until the author's death in 1939. It would be rash of me to claim that no commentary already exists, since the dream book has many times been combed for its autobiographical insights, for its story of the discovery of psychoanalysis, for its introduction to the methods and theory of analysis, and for its dreams—most of which have been subjected to enthusiastic reinterpretations by other persons. Yet I know of no concerted attempt to examine the book critically so as to take into account both the construction of the argument and Freud's marvelous self-presentation.

Briefly stated, here is what I argue: that as its title suggests, Freud's book is about interpretation, but that it never would have been as persuasive without its pretension to science; that the procedures of the book are inductive up to a point, but the arbitrary turnings that Freud takes can best be explained as wishful, pleasing not only to the writer but to his readers; that his interest and that of his readers in secrets and detection has a history (which cannot be traced here but has been touched upon in my other work); that Freud's analysis of his dreams of ambition always falls within polite bounds and steadily redounds to his credit; that his modest ambition is also a product of its time and may usefully be seen as inspiring the Oedipus complex rather than the other way around; that the institution of the censorship, very much a feature of the dream book, is also and necessarily socially grounded; that the genre of the book, considered as a frame story embracing many dreams and anecdotes, is a romance or serial comedy; that Freud occasionally employs make-believe—that is, invents evidence, if one insists on thinking of the book as science—but generally within the parameters of a conventional homiletics; and that the finale of the dream book, and much of the humor throughout, can be understood only as a particular kind of fine performance. *The Interpretation of Dreams* is in some respects an interpretation of ambition; it turns ambition, understood as a set of motives to be acted upon, into wishfulness, a varied and nearly boundless set of aggressive fantasies more suited to storytelling.

One reason for such a commentary is simply practical. Freud's influence on the intellectual life of the second half of the twentieth century in the West is perhaps greater than that of any other writer of his time, so it becomes impossible, short of taking a lifetime or forming a committee, to examine very seriously more than one text at a time. Sebastiano Timpanaro's commentary on *The Psychopathology of Everyday Life* may be thought of as a model for the present one on the dream book. Unfortunately, more expert attention has been directed to the outline of Freud's thinking in his private correspondence and papers than to the manner in which it became fixed by public presentation in *The Interpretation of Dreams*. Like Timpanaro and others, I have become very concerned with the arguments to which the writer thus committed himself, the more so because he typically reveals that he was aware of their shortcomings. Even if one discounts the relation of psychoanalysis to science, these arguments need to be examined much more closely if we expect to gain any just understanding of Freud's influence on modes of interpretation in literary, historical, and social studies today—to say nothing of popular psychology. The present commentary frankly emphasizes the pleasures of the dream book, its accommodations and appeal, in an attempt to account for its undoubted success. I also stress the dependence of its arguments on historical conditions and therefore call for more study of those conditions. At the close of the commentary, I try to indicate where I stand with respect to the current criticism of psychoanalysis.

"*The Interpretation of Dreams* was a revolutionary work," Frank Kermode has recently remarked; "it did not affect general ideas about interpretation at once, but over the century its influence has been decisive." To a certain extent, while not forgetting the book's place in the development of Freud's thought and its wide dissemination, I have tried to re-create the experience of reading it for the first time. This initial effort to detach the argument from received ideas about it was made possible by the participation of very bright students, all of whom knew something about Freud but none of whom had ever read the dream book, in seminars over the past few years. The rule followed in those seminars was the same as that of the commentary: never to dismiss any turn in the argument as unwarranted or the inclusion of any illustration as irrelevant without trying to define what new ground the unexpected turn achieved or what purpose the illustration finally served. Besides other writings by and about Freud, the seminarians pored over his favorite Dickens novel, *David Copperfield,* as a control exercise in the study of fiction and wish fulfillment—an exercise that may explain the mysterious surfacings of that novel in the pages that follow.

My greatest debt is to the UCLA and Yale students who faithfully read the dream book, page by page, and made plain their reactions to it. I could scarcely have found a more diverse group of students, in terms of ethnic background, economic well-being, or academic field, if I had hired an opinion pollster to select them: not only their shared intelligence but their varied perspectives benefited us all. Meanwhile a number of people, including J. William Schopf, Helene Moglen, and Nina Auerbach, prompted some friendly occasions for trying out my own conclusions about Freud's censorship. Two colleagues professionally committed to psychoanalysis, Herbert Morris and Peter Gay, generously encouraged me in this effort to breach disciplines, regardless of their partial or total disagreement with what I was saying. Ruth Bernard Yeazell, the best reader one could dream of, was immensely helpful with my second draft, as always. So did the readers for Princeton University Press, and the diplomatic Robert Brown, help substantially, after the final draft was expertly retyped by Diane Repak. To all such readers and listeners, my repeated thanks; and of course I accept responsibility for any errors of fact or judgment that remain.

Virtually all studies of Freud in English are profoundly indebted to the Standard Edition, the labor of James Strachey and his colleagues, published in London by Hogarth Press and the Institute of Psychoanalysis and at present distributed in the United States by W. W. Norton. I think it foolish to carp—as some do—about the translation of this edition, which is a wonder of completeness, cross referencing, and bibliographical exactitude. When quoting from it, I have regularly consulted the *Gesammelte Werke*; and where translation has seemed to close down some alternative meaning, or I think my readers might wish it, I have supplied the original German in parentheses. At many points I am indebted to various authorities on Freud and have indicated so by naming them and citing the reference. Perhaps footnotes could have expressed my agreement or differences more exactly, but for a commentary on a book that—starting with Freud himself—has been more often footnoted than carefully reconsidered, I decided to rule out discursive notes and write of the dream book as simply as possible.

Freud's Wishful Dream Book

CHAPTER ONE

"A Dream Is the Fulfilment of a Wish"

*T*HE INTERPRETATION OF DREAMS deserves to be regarded as Freud's masterpiece, in two widely accepted meanings of that term: a famous work that contains many of his most
important ideas, and the work that qualified him not merely as
a member but as the founder of a guild. Among those most devoted to the study of psychoanalysis, both insiders and outsiders
have affirmed the book's prominence. The early disciple and
official biographer Ernest Jones calls it "Freud's most original
work," which established "a secure basis for the theory of the
unconscious in man" and remains "the best known and most
widely read" of his contributions (1953, 1.350–51). Peter Gay, historian and biographer, concedes that "the genre of Freud's masterpiece is . . . undefinable" but fixes its position with a military
metaphor: "In the evolution of Freud's psychoanalytic thinking,
The Interpretation of Dreams occupies the strategic center, and he
knew it" (1988, 104, 117). In a philosophical assessment, Richard
Wollheim names it Freud's "most important work" and "also a
work of confession" (1971, 60); and Frank J. Sulloway agrees,
from a perspective in history of science, that the same "has generally been considered Freud's single most important book"
(1979, 320), though he opposes the legend that grew up around
it.

Freud was forty-four when he completed his masterpiece and
seventy-five when he declared, in the preface to the third English

edition, that "it contains, even according to my present-judge-ment, the most valuable of all the discoveries it has been my good fortune to make. Insight such as this falls to one's lot but once in a lifetime." Therefore he can be said to have anticipated later assessments of the work, though it is not quite certain which discoveries he was referring to here. Presumably he meant the schema of the mind set forth in the final theoretical chapter, including the primary and secondary processes, repression and the crucial borderline between unconscious and preconscious thoughts.

> The two psychical systems, the censorship upon passage from one of them to the other, the inhibition and overlay-ing of one activity by the other, the relations of both of them to consciousness—or whatever more correct interpre-tations of the observed facts may take their place—all of these form part of the normal structure of our mental in-strument, and dreams show us one of the paths leading to an understanding of its structure. If we restrict ourselves to the minimum of new knowledge which has been estab-lished with certainty, we can still say this of dreams: they have proved that *what is suppressed continues to exist in nor-mal people as well as abnormal, and remains capable of psychi-cal functioning.*

That these words in chapter 7 provide a formal conclusion Freud suggests by repeating from his title page a motto from Virgil, "Flectere si nequeo superos, Acheronta movebo," and by adding as a separate paragraph in 1909, "*The interpretation of dreams is the royal road to a knowledge of the unconscious activities of the mind*" (607–8).* Earlier in the chapter he has apologized for fall-ing back on his clinical experience "instead of proceeding, as I should have wished, in the contrary direction of using dreams as a means of approach to the psychology of the neuroses" (588).

* Whenever I cite page numbers alone, without a date, the reference is to *The Interpretation of Dreams* in the standard edition of Freud's works (1953–74). For this and all other references, see the index to works cited, at the back of the book.

Thus we learn that the dream book does not stand entirely on its own but works in tandem with Freud's studies of hysteria in the last decade of the nineteenth century. Since 1950 we have known that it also had a direct, sequential relation to the "Project for a Scientific Psychology" draft of 1895, which provided an outline for the psychology in the theoretical chapter. Freud's masterpiece can even be regarded as the completion of, or alternative to, the abandoned project, since the last few sections of the draft identified dreams with wish fulfillment and sketched a paradigm of the dream work (1950, 335–43). In short, *Die Traumdeutung* begins where the project left off. Though Freud's title would lead one to expect a hermeneutics, he continues to refer to his method as scientific. The long first chapter, written last, is a review of "The Scientific Literature" on dreams. Thereafter he explicitly identifies interpretation with "popular procedures" and his own method with science, though to be sure he seems to want it both ways: "I must affirm that dreams really have a meaning and that a scientific procedure for interpreting them is possible" (99–100).

The book has an argument, all right, which gets under way as early as the first chapter, since Freud's opinion of previous writers to some extent betrays his position in advance. The real beginning is the second chapter, introducing the method by way of "An Analysis of a Specimen Dream"—the dream subsequently referred to as the dream of Irma's injection. Thus Freud adopts a common and effective mode of exposition, by offering an illustration of his approach before classifying dreams and ascribing rules for them. Moreover, since the dream of Irma's injection was the first dream he "submitted to a detailed interpretation" (106n), the exposition is of the kind that shares with its readers a narrative of discovery. The book offers a story of the investigation, an attractive and often persuasive story because it takes readers into the investigator's confidence. The third chapter—the only one with a proposition as the title—states a clear thesis, "A Dream Is the Fulfilment of a Wish." Another short chapter then introduces "Distortion in Dreams," distortion already apparent between the manifest and latent content of the specimen dream. The fifth, "The Material and Sources of Dreams," promises data to work on; and the sixth and longest chapter, "The Dream-Work," a breakdown of the different forms of distortion that occur. The seventh is the theoretical chapter already mentioned, "The Psychology of the Dream Processes," which describes much more abstractly a mechanism designed to accommodate these findings.

The formal order of *The Interpretation of Dreams*, however, is nothing like the experience of reading the book. Readers are likely to be impressed at one moment, annoyed at the next, and—in all but the first and last chapters, perhaps—genuinely entertained. Some reaches of the text, like the section on "Representation by Symbols in Dreams" in chapter 6, which did not exist at all in the first edition and enlists a method of interpretation that Freud initially disclaimed (97), are unintentionally comic; others, like the "Dreams of the Death of Persons of Whom the Dreamer is Fond" section in chapter 5, are suffused with such thoughtfulness that they might belong to another text altogether. It has to be said that—for a careful reader who hopes to find evidence and argument in substantial agreement—some of the more breathtaking moments, among so much that is impressive or entertaining, are unexpected lapses or very long leaps of inference. The exposition of a theory of dreams and their interpretation is compromised throughout by other sorts of interest and ordering, which are those of a frame story—in this case, a collection of dreams and anecdotes of generally comic import, some of which pertain to the narrative of discovery more than others. Another explanation of the number of surprises and shifts in the argument, however, is that Freud was pretty much improvising as he went along. At any rate he left signs of improvisation in the text that he published, and for subsequent editions he revised the text without ever removing these. Rhetorically this was not necessarily a mistake, since Freud appears to share with his readers the uncertainty of discovery.

To get a fair idea of what the experience of reading the dream book is like, consider the argument that follows upon the analysis of the specimen dream in chapter 2, which ends, "*When the work of interpretation has been completed, we perceive that a dream is the fulfilment of a wish*" (121). The next chapter, which bears the last clause as its title, begins with a charming paragraph depositing readers and author alike (subsumed by the first-person plural) upon "high ground" amid "the finest prospects," and "in the full daylight of a sudden discovery" (122). This is all very encouraging, as is the series of tough questions posed for the theory in the next paragraph. Then there is a line break, and a proposal to "leave all these questions on one side" while we inquire, reasonably enough, whether wish fulfillment is "a universal characteristic of dreams" and not merely the case in respect to the dream of Irma's injection (123). Beyond this point, however, the new chapter with its bold assertion of a thesis becomes disappointing. "We," as a joint enterprise and investigating team, are forgotten, and readers are offered summary versions of approximately thirteen dreams that, sure enough, but rather lamely, express some fulfillment of a wish. Thirteen plus one instances do not a universal proposition make. What seems to happen is that the anthologizing tendency of the frame story takes over here, without more concern for inquiry or logic. Freud does manage to advance his theory a little in this chapter, by suggesting that wish fulfillment will be found in a simpler form in the dreams of children and by casually referring in his last paragraph to "the *hidden* meaning of dreams" (132, my emphasis). At the very least, the enthusiasm and rigor of the first few paragraphs and the generally pleasing, if bland, account of new dreams seem to make it now too late, or churlish, to question the analysis of the specimen dream that has gone before.

Freud apparently senses that readers are not likely to be happy with chapter 3, since before anyone has had the chance to read it and protest, he commences chapter 4, not charmingly but testily: "If I proceed to put forward the assertion that the meaning of *every* dream is the fulfilment of a wish, that is to say that there cannot be any dreams but wishful dreams, I feel certain in advance that I shall meet with the most categorical contradiction" (134). Typically, Freud does not accept any new cards but ups the ante. At the same time, he grasps the logic of his contention a little more firmly this time, by admitting that anxiety dreams pose a problem. To a gambler in ideas, a problem may be an opportunity: if he can show that even dreams that convey the opposite sensation nonetheless express wish fulfillment, he can begin to defy contradiction and thereby strengthen the impression that his proposition will hold across the board. Yet Freud puts this argument aside also, while he introduces another consideration: "There is no great difficulty in meeting these apparently conclusive objections. It is only necessary to take notice of the fact that my theory is not based on a consideration of the manifest content of dreams but refers to the thoughts which are shown by the work of interpretation to lie behind dreams." Instead of completing the argument, in short, he shifts the evidentiary ground. "We must make a contrast between the *manifest* and the *latent* content of dreams" (135); and it is of course in the latent content, the content revealed by interpretation, that he will look for evidence of wish fulfillment. This shift is obscured somewhat by a promise to pursue both lines of argument together, because—in a homely simile—"it is easier to crack two nuts together than each separately." Thus he proposes to add to the question of "How can distressing dreams and anxiety-dreams be wish-fulfilments?" the question "Why is it that dreams with an indifferent content, which turn out to be wish-fulfilments, do not express their meaning undisguised?" (136). But of course he addresses only the new question, which conveniently contains within it the new evidentiary ground (dreams that "turn out to be wish-fulfilments") and a logic of its own (a dream that seems to tell one story but originally told another has been disguised). Hence this chapter, which importantly introduces the idea of censorship, is given the title "Distortion in Dreams" (Traumentstellung).

A serious consequence of Freud's raising the question of anxiety dreams in chapter 4 only to put it aside is that he never quite confronts the possibility of there being more than one general etiology for dreams until *Beyond the Pleasure Principle* (1920, 13–14, 32–33). A good many of the dreams Freud recounts in 1900 may seem to his readers to convey fear or anxiety, but neither in chapter 4 nor later does he regard any of these as challenging his present theory. In his *Introductory Lectures on Psychoanalysis* the approach to dreams changes somewhat but not, he makes clear, to the point where he could be said to make any "concession." On the contrary, it has become "our duty . . . to be able to indicate the wish-fulfilment in any distorted dream we may come across, and we shall certainly not evade the task" (1916–17, 219). Another consequence of his proposal to crack two nuts together is that the second question is never examined very closely. Thus Freud begins to assume that all dreams are disguised, even though he has just mentioned some in the previous chapter that are not disguised and will later observe that a few dreams of particular interest to him represent *un*disguised wishes for the death of relatives (266–67). Notwithstanding the importance of distortion to Freud's argument, and his introduction of the censorship, his intention to seek out dreams that would contradict the theory is frittered away entirely when he descends to telling how patients have dreams whose latent wish is to prove they are not wishful—that is, to prove Freud wrong. The chapter concludes, "We shall be taking into account everything that has been brought to light by our analysis of unpleasurable dreams if we make the following modification in the formula in which we have sought to express the nature of dreams: *a dream is a (disguised) fulfilment of a (suppressed or repressed) wish*" (160).

This formula is much closer to what Freud believes than the first. The conditions admitted parenthetically are the true objects of his investigation, yet enclosing the words in parentheses suggests that the underlying thesis has not changed. The chapter on distortion is pivotal in his argument, as Freud appears to be working out the argument as he proceeds (and this was very probably the case). Therefore his readers are privileged to follow his thinking, while at the same time he has been right from the first: a dream is still the fulfillment of a wish. The claim that this proposition is universal, though its proof keeps getting put off, is impressive. But there is a downside to attractive self-confidence. If Freud could not admit any significant exceptions to the thesis until 1920, apparently he was taken in for twenty years by his own rhetoric. For skeptical readers, neither the claims to universality nor the existence of anxiety dreams nor matters of disguise have been satisfactorily resolved as yet.

Among the skeptics, though far from unsympathetic to psychoanalysis, was Ludwig Wittgenstein. In conversations with Rush Rhees four years after Freud's death, Wittgenstein protested that "it seems muddled to say that *all* dreams are hallucinated wish fulfilments" (1966, 47). Note that he is careful to specify "hallucinated." Perhaps the most remarkable difference between dreams and diurnal perception is this one, that of hallucination. In almost the last words of the draft project and in chapter 7 of *The Interpretation of Dreams*, Freud was mindful of the difference as well, but for most of the book his language simply takes it for granted. Readers understand that this is a form of shorthand, yet simplification undoubtedly makes the thesis more appealing: a dream is the fulfillment of a wish. Only if dreams came true, as we say, could this proposition strictly make sense. Whereas some wishes, conscious or unconscious, are actually carried out, dreams present only a story or a scene, whether suggestive of the fulfillment of a wish or not. This is perhaps the main reason that, to demonstrate the power of unconscious wishes, Freud needed also to write *The Psychopathology of Everyday Life*. Wittgenstein provides this reminder about the hallucinatory nature of dreams and further remarks that to insist that they are all wish fulfillments seems muddled. He does not meet Freud's universal with a "most categorical contradiction," as predicted, but does suggest why it is muddled:

Partly because this doesn't seem to fit with dreams that spring from fear rather than from longing. Partly because the majority of dreams Freud considers have to be regarded as *camouflaged* wish fulfilments; and in this case they simply don't fulfil the wish. Ex hypothesi the wish is not allowed to be fulfilled, and something else is hallucinated instead. If the wish is cheated in this way, then the dream can hardly be called a fulfilment of it. (1966, 47)

It is as if he had been rereading chapter 4, since he addresses the same two questions, of anxiety dreams and disguised meanings, with which that chapter begins.

These comments from Wittgenstein seem just, and they should be amplified. Observing the strain placed on logic in *The Interpretation of Dreams* is a first step in learning how to read the book. Very typically Freud will begin with a set of phenomena—such as dreams in general—and then pass on to a narrower subset with a particular characteristic, which he then writes about as if it pertained to the larger set. Thus dreams immediately become wish dreams, and wish dreams then become dreams with a disguised or hidden wish, and these dreams soon become guilty dreams—for the purport of section A of chapter 5, "The Material and Sources of Dreams," is to show that there is no such thing as an "innocent" dream (183). Freud's order of exposition seems scientific, because it mimics the narrowing down of a search for causes of dreams, whereas all that is really happening is that subsets of dreams are narrowing to those that most interest him. Yet he persists in referring to dreams in general, without qualification: "Nothing that has *really* remained indifferent can be reproduced in a dream" (182), for example. Though he contends that there are no dreams except those that respond to a wish, he does not bother to contend that, within the next subset, there are no dreams but those that respond to disguised wishes. Yet he has even given a few examples of undisguised wish dreams, mostly of children (127–31). The argument waives those dreams, and readers are encouraged to forget them. The dreams left behind figure like rejected causes in a true inductive process; but in fact nothing has been ruled out, merely discarded. Similarly Freud claims (279), and his method generally implies, that a dream may have multiple interpretations; but he nearly always presents his interpretation as the only one. His search for clues may resemble a criminal investigation, but this investigator seldom questions whether he is in possession of the only explanation of the facts. Whether as science or as any other use of induction, the arguments pursued by Freud are not very satisfactory.

Wittgenstein believed such arguments should be characterized as speculation. "Freud is constantly claiming to be scientific. But what he gives is *speculation*—something prior even to the formulation of an hypothesis" (1966, 44). Freud might agree that his thinking is of the kind that precedes the definite formulation of a hypothesis, because he often professes a readiness to change his view as more facts become known. Too commonly, however, his speculation rapidly stiffens into an article of faith. In the dream book, for example, a query to himself whether thoughts reaching back to childhood may not be a "precondition" for dreaming (218) later becomes an assertion that, although it "cannot be proved," cannot be *dis*proved (554) and eventually emerges as "one of the corner-stones of our teaching" (589). Because Freud lays this cornerstone in place gradually, he appears to be proceeding thoughtfully; but the idea that all dreams have some infantile motive remains nearly as fanciful as it was at the beginning, since it is ignored by most of the dreams Freud himself analyzes. By speculation, Wittgenstein points to "the sort of explanation we are inclined to accept," or an explanation with "the attraction which mythological explanations have" (1966, 43). More needs to be said about the attractiveness of Freud's explanations, for it is their very claim to science that constructs the myth. One has only to ask how effective a book like *The Interpretation of Dreams* would be without this claim, or without the array of anecdotal history and supposed inductive truths. Would speculation presented *as* speculation have nearly the same force? This question raises the disturbing thought that only insofar as Freud keeps up a vigorous pretense of science (pretending to himself, no doubt, as well as to his readers, and always with great rhetorical skill) can his theory be remarkable, and the corollary that only as we accept the pretense can we be very much moved.

The power of this double appeal should not be discounted. If Freud was to be a prophet of his time, he could hardly pretend to anything *but* science. In assessing this prophetic role, Harold Bloom adopts Wittgenstein's term *speculation:* among "a multitude of changes in the modes of interpretation available to the West," according to Bloom, "the Freudian speculation has perhaps been most influential in our century, if only because we now find it difficult to recall that psychoanalysis, after all, is only a speculation, rather than a science, a philosophy, or even a religion." To underscore his view, Bloom adds, "Freud is closer to Proust than to Einstein, closer even to Kafka than to the scientism of Darwin" (1989, 146). Though I share this view far more than any that persists in believing the procedure reported in the dream book to be scientific, the name that leaps out here is Darwin's. Surely Darwin was a prophet before Freud, and his science was fully necessary to that role.

It is worth digressing a moment to insist on Darwin's inspiration for Freud, on two grounds alone: the theory of natural selection had been launched by a single book, made accessible to a wide public; and Darwin's inductive logic was extraordinarily daring, surmounting vast lacunae in the evidence and many needed explanations that had not as yet occurred to the author or anyone else. We also know that Freud subsequently compared his achievements to Darwin's (1917, 140–41; 1916–17, 285). His specific allusions were mostly to *The Descent of Man,* but the tremendous influence of *On the Origin of Species* should not be forgotten. There was an obvious thematic similarity in the thinking of the two men; both can be said to have been concerned with descent, whether of a species or the individual. "By my theory," Darwin had written, "these allied species have descended from a common parent; and during the process of modification, each . . . has supplanted and exterminated its original parent and all the transitional varieties between its past and present state" (1859, 173). In many respects the theorizing of Darwin and Freud was continuous, and the substantial relation of the two has been traced by Sulloway (1979) and Lucille B. Ritvo (1990). The founder of psychoanalysis picked up a good many rhetorical techniques from *On the Origin of Species* as well. Darwin never failed to express surprise at his own temerity, for example, or to confess the inadequacy of the present state of his science. "Our ignorance of the laws of variation is profound," he would typically admit, and forge relentlessly onward with his argument. Like Freud, he anticipated his readers' doubts: "Long before having arrived at this part of my work, a crowd of difficulties will have occurred to that reader. Some of them are so grave that to this day I can never reflect on them without being staggered; but to the best of my judgment, the greater number are only apparent, and those that are real are not, I think, fatal to my theory" (1859, 167, 171). The difference, I suggest, is that Darwin openly and ardently pursued these doubts home without beating about the bush. In one respect he was merely being polite: *more* difficulties have occurred to him than to most readers, and those two sentences introduce three chapters of measured response to difficulties raised thus far by his theory. The whole book, or "one long argument" (1859, 459), was a reply to objections and a review of alternatives—and appropriately so, because the way to show that the evidence could be accounted for by natural selection was to show that it could be accounted for in no other likely way. Darwin had the advantage of being able to refute one principal set of received opinions—that species were separately created and immutable—and he took great pains to limit his argument and render it acceptable. Freud's masterpiece may be more pleasurable, and especially valuable for its mythography and autobiography, but Darwin accepted the full burdens of reasoning from effects to causes.

When Wittgenstein returned to the subject with Rush Rhees in 1946, he complained, "I have been going through Freud's 'Interpretation of Dreams' with H. And it has made me feel how much this whole way of thinking wants combatting" (1966, 50). One aspect of the thinking that obviously bothered him was the use of free association to arrive at determinate causes. The capacity of free association to shape itself into a plausible narrative seems to be mainly a function of time and energy.

> The fact is that whenever you are preoccupied with something, with some trouble or with some problem which is a big thing in your life—as sex is, for instance—then no matter where you start from, the association will lead finally and inevitably back to that same theme. Freud remarks on how, after the analysis of it, the dream appears very logical. And of course it does.
>
> You could start with any of the objects on this table—which certainly are not put there through your dream activity—and you could find that they all could be connected in a pattern like that; and the pattern would be logical in the same way. (1966, 50–51)

In other words, free association cuts too wide a swath and is too creative not to come up with some grains of coherent narrative, and even though the emerging narrative is likely to be important to the person engaged in the analysis (whether the patient or the analyst), there can be no guarantee that it is the cause of the dream unless there is no other likely cause. Philosophers who essentially repeat Wittgenstein's arguments have emphasized that it is precisely Freud's method of analysis that the dream book is supposed to demonstrate. Clark Glymour concludes that the method is "worthless": even Freud's frequent appeal to "two or more elements of the dream which are independently associated with a key figure in the interpretation" yields only the appearance of a connection (1983, 65, 67). Adolph Grünbaum makes a similar argument by focusing on the specimen dream in chapter 2, for which Freud's interpretation fails to show how free association can recover "a *repressed* thought" and thereby fails to authenticate the method as a whole, especially not "his *substantive* theory of dreams as *universally* wish fulfilling" (1984, 222, 224). But of course the method cannot be worthless, for as Wittgenstein suggests, it yields "the sort of explanation we are inclined to accept." At a certain point, even without waiting for a lengthy course of free associations, an analyst or a person who has had a dream decides that such was the explanation—the reason for it, or "the cause" in a colloquial sense. Such reasons, when specific to the person and the circumstances, are more satisfying to most people than scientific causes: the explanation serves as a personally relevant myth. Even when explanations are codified, as in Freud's subsequent surrender to dream symbolism or his determination that the "motive force" of the dream must originate in infancy (589), some people value these explanations for their predictability or their confirmation of some article of faith, such as that all dreams are sexual or return one to childhood.

What seems to worry Wittgenstein, or what stirs him to say that "this way of thinking wants combatting," is the way such explanations pose as something other than what they are. "Analysis is likely to harm," he says. "Because although one may discover in the course of it various things about oneself, one must have a very strong and keen and persistent criticism in order to recognize and see through the mythology that is offered or imposed on one. There is an inducement to say, 'Yes, of course, it must be like that.' A powerful mythology" (1966, 51–52). It is important, then, "to recognize and see through" whatever fictitious constructions may pose as facts. Freud often seems to do all he can to prevent even himself from performing this operation successfully, mainly by asserting claims to science. It may be that, in an age of science, his speculations might have taken this form even if he had not been trained in neurology. The long chapter on the dream work, in which can be found some of Freud's most durable thoughts, begins with claims of precedence such as scientists dream of:

> Every attempt that has hitherto been made to solve the problem of dreams has dealt directly with their *manifest* content as it is presented in our memory. All such attempts have endeavoured to arrive at an interpretation of dreams

from their manifest content or (if no interpretation was attempted) to form a judgement as to their nature on the
basis of that same manifest content. We are alone in taking
something else into account. We have introduced a new
class of psychical material between the manifest content of
dreams and the conclusions of our enquiry: namely, their
latent content, or (as we say) the "dream-thoughts" arrived
at by means of our procedure. It is from these dream-
thoughts and not from a dream's manifest content that we
disentangle its meaning. (277)

As myth, certainly, these claims have succeeded in justifying
themselves. The preamble to the chapter continues with a now
famous analogy between these two representations of a dream
and "two different languages," one "as it were in a pictographic
script" and the second its translation—the analogy to which
Jacques Lacan (1956) and Jacques Derrida (1966) have given
currency.

In stating these claims, Freud seems to forget that he belatedly
introduced the distinction between the manifest and latent content of dreams in chapter 4 as a way of putting on hold the still
untested theory that all dreams conveyed the fulfillment of a
wish. If the dream thoughts, or hidden meaning of the dream,
could be shown always to answer a wish, then the general theory
would hold even though some dreams manifested themselves as
anxious or otherwise unpleasurable. Now, introducing his observations on the rules governing the translation of dream thoughts
as the dream, he boldly treats inferences from the dream as evidentiary fact, part of the data on which his theory is based. Here
and throughout the book the dream thoughts are, and can be
nothing but, the narrative supplied by Freud's interpretation.
Each such narrative, if it has any relevance to the dream at all,
must be inferred from some of its details or the whole, together
with any circumstances that may be relevant. To say that "we
have introduced a new class of psychical material," the latent
content or dream thoughts, ought not to be registered as a boast

but as an admission. The new narrative must be derived from the dream that became its expression. So persuaded is Freud that he can always supply this new narrative that the dream thoughts have become, always already, facts in evidence. The circularity in the use of evidence appears everywhere in the book and, in short, wants combating.

Instances of circularity range from petty confusions of a dream with its interpretation to contentions that are finally self-defeating in Freud's attribution of motive force to the unconscious. To cite a petty instance: a portion of one of his patient's dreams, he says, "reminded me, when I heard it, of the masterly introduction to Alphonse Daudet's *Sappho*," though the dreamer tells Freud that it reminded *him* of a play he saw the night before (285–86). Freud nevertheless continues for a couple of pages to deploy associations with Daudet in interpreting the dream, which he subsequently refers to as "my patient's *Sappho* dream" (305). In yet a third passage on this dream, "the introductory scene from Daudet's *Sappho*" is said to supply the "prototype in the dream-thoughts" (326). One does not worry so much about the patient as about Freud's willingness to deceive himself. The same carelessness about where the evidence is coming from can have larger consequences. In a telling observation about his own dreams, Freud writes, "A dream is in general poorer in affect than the psychical material from the manipulation of which it has proceeded. . . . The dream work has reduced to a level of indifference not only the content but often the emotional tone of my thoughts as well. It might be said that the dream work brings about a *suppression of affects*." And he ponders "the incontestable fact that large numbers of dreams appear to be indifferent, whereas it is never possible to enter into the dream-thoughts without being deeply moved" (467). Now the dream thoughts are the product of Freud's interpretation of his dreams, and interpretation, even if it succeeds in picking up some thoughts from the preconscious, is a conscious process. Therefore this argument is a tribute to the far greater affect provided by his own narrative powers, by which he is always "deeply moved." Freud does not stop to think of the implications of such circular appeals to evidence, which are bound to be self-defeating if dream interpretation is truly the road to a knowledge of the unconscious.

Different readers will respond differently to the gerrymandering of evidence, to the presumption of a doctor-patient relation that turns disagreement into resistance, or to the pace with which a metaphor—most notably the censorship in this book— can become an institution with a life of its own. I am particularly distressed by Freud's lapses or slides in the argument but also by his failure to follow rules that he sets down for himself. He begins, for example, by positing the need to know the context of a dream, a requirement that is said to limit the author to his own dreams or those of his patients (104); but before the anthology of dreams and interpretations is at an end this rule is stretched further and further, until the last chapter features a father's dream of a dead child that "was told to me by a woman patient who had herself heard it in a lecture on dreams" (509), hearsay that Freud treats just as seriously as he treats his own dreams. Nicholas Rand and Maria Torok have recently characterized Freud's thinking as "divided and self-contradictory," since he creates such a "fundamental methodological discrepancy" as that in the dream book between the rule of free association that he has set down for himself and the recourse notwithstanding to stable symbolic meanings. Thus they worry that "Freudian theory fractures at its very core" (1993, 568–70).

One of the earliest disturbing turns in *The Interpretation of Dreams* occurs in chapter 2 after a brief discussion of free association and before the specimen dream, when he discounts the use of patients' dreams as impractical. "Thus it comes about that I am led to my own dreams, which offer a copious and convenient material, derived from an approximately normal person and relating to multifarious occasions of daily life" (105). Yet neither here nor in the preface to the first edition does he pause over the question whether the dreams of a single person can be any more representative than those of his patients. Rather, he expands a little huffily on the risks to himself of possible indiscretions. As Freud is perfectly aware, his readers are in no danger from that risk. On the contrary, their curiosity will be piqued. After presenting the specimen dream and its interpretation, he returns almost truculently to the same theme: "If anyone should feel tempted to express a hasty condemnation of my reticence, I would advise him to make the experiment of being franker than I am" (121; see also 136).

The dream of Irma's injection is the only one in the entire book for which an exact date is given. Freud wants thereby to celebrate the occasion, but since his presentation is also uniquely formal it may be that it is the only dream *cum* interpretation for which he has adequate notes. The business about his reticence and the concealment of some of his thoughts, with the air of defiance at the end, give rise to the possibility of still deeper games being played. Obviously Freud was frank enough to concede that he was not being completely frank. Even as the dream's meaning "was borne in on me," he writes, he experienced "difficulty in keeping at bay all the ideas which were bound to be provoked by a comparison between the content of the dream and the concealed thoughts lying behind it" (118). So it is not possible to draw a clear line between relevant facts that Freud has consciously altered or withheld and thoughts not fully realized by himself, such as he refers to mystically as the dream's "navel" or "point of contact with the unknown" (111n). Remember that the dream of Irma's injection figures in the book Freud is writing as a specimen dream with its interpretation, and ever since as *the* specimen dream of psychoanalysis. But "Irma's injection"—with quotation marks, because of course "Irma" has been substituted for the name of the patient in the dream and any "injection" is merely a conjecture of the dreamer—figured in Freud's life as the occasion of his discovery, the method that revealed to him for the first time the meaning of dreams and sealed the invention of psychoanalysis. Hence, six months after publication of his masterpiece, while staying again in the house in Bellevue outside Vienna, where he had been at the time, Freud wondered in a letter to his friend Wilhelm Fliess (1985, 417) if a marble tablet would ever be raised there, with the inscription:

> Here, on July 24, 1895,
> the secret of the dream
> revealed itself to Dr. Sigm. Freud

James Strachey's annotation of the text in the standard edition also provides this inscription and thus enables readers to share in the historical significance of the dream even while reviewing Freud's method, which can be traced in the difference presented by two narratives—first a verbal transcript of the dream as remembered and then the dreamer's associations with one segment after another.

Though the initial scene is a social one, in which "we" are receiving people and Irma is one of the guests, the action of the specimen dream can be summarized as a perplexed and somewhat anxious examination of this female patient by four male doctors—one consultant referred to as Dr. M., two friends referred to as Otto and Leopold, and the dreamer himself. The injection is not given to Irma in the present action but recalled near the end as having been given her recently by Otto, possibly with a dirty syringe. In describing the circumstances of the dream, Freud has already explained that he felt Otto was blaming him for Irma's failure to recover fully. His transcript of the dream, about 350 words, contains many details that are not immediately explicable. After reviewing, in about 4,700 words, the various facts and thoughts he has succeeded in recalling by association with the dream, however, he more or less hits on the last thought in the transcript itself to find his wish: "The conclusion of the dream, that is to say, was that I was not responsible for the persistence of Irma's pains, but that Otto was" (118). By running down the obscure details of the dream, Freud's interpretation enacts much useful introspection and creates the first autobiographical interest of the dream book, but the moral of the story can be deduced by most readers from the dream's denouement without this information. Even the working title he assigned to it—"the dream of Irma's injection"—features the end of the manifest dream as scripted.

Here and elsewhere Freud adopts the convention of personal memoir writing, often imitated by novelists, of substituting initials and fictitious names for some of the actors, while generally he appears without disguise under the first-person pronouns. The same convention is sustained by the interpretation, though this first interpretation refers directly to his wife and eldest daughter and unmistakably, though without naming him, to his friend Fliess. Not surprisingly, the dream of Irma's injection has become—perhaps always was—a roman à clef, to be retold and reinterpreted many times by Freud's followers. The most sensitive reinterpretation may still be that of Erik Erikson (1954), whose sympathetic and wide-ranging reflections command respect. To be sure, Grünbaum (1984, 225–30) makes short work of Erikson's attempt to supply a motivating infantile wish for the dream. But Erikson deserves credit for anticipating Grünbaum's own principal criticism, that getting back at Otto in the dream is a thought far too near consciousness to be considered a *repressed* wish. He also deserves credit for imaginatively sketching Freud's early years and the milieu of the dream and its interpretation, a line of thinking that anticipates the treatment of the dream book as social history by Carl Schorske (1973) and William J. McGrath (1986).

Erikson did not know as much about "Irma's injection" as present-day commentators, however, because fewer of Freud's letters to Fliess were then available. When Marie Bonaparte, Anna Freud, and Ernst Kris published *The Origins of Psychoanalysis* (1950), they omitted anything that might detract from Freud's memory or diminish their own idea of him. Max Schur's paper on the specimen dream (1966) first made public the letters pertaining to the unfortunate Emma Eckstein, a patient whom Freud was treating for hysteria. A century later, Freud seems to have been much too trusting—not to say sanguine—in subjecting his patient to Fliess's surgery. But Fliess held that hysterical symptoms might be relieved by surgery on the nose (his specialty) because of that organ's hidden physiology—in the words of Freud's dream analysis, his friend "had drawn scientific attention to some very remarkable connections between the turbinal bones and the female organs of sex" (117). (Only a surgeon could conceive of such Shandean matter as female, perhaps, though Sulloway (1979, 135–237) helps to put all this in historical perspective.) Fliess had operated on Freud himself and was generally advising him as to his health, while Freud also counted on him as a sounding board for his own theories. In February 1895 Fliess operated on Eckstein in Vienna and returned immediately to Berlin. But her wound did not heal. Freud called in another specialist, who was able to demonstrate what the trouble was by pulling out a half meter of gauze that the surgeon had inadvertently left inside the patient, who now lost consciousness along with a large quantity of blood. The blood and smell—and perhaps more pertinently, "strong emotions"—drove Freud from the room, a retreat that Eckstein had the good grace to kid him about. All this has become well known: Freud's report to Fliess on 8 March 1895 makes stunning reading, historical documentation of high drama between two friends (1985, 116–18). Freud apologizes, in effect, for getting his colleague into such a mess: Fliess cannot be to blame. And Freud carries on in this vein for over a year: "I shall be able to prove to you that you were right, that her episodes of bleeding were hysterical, were occasioned by *longing*, and probably occurred at the sexually relevant times (the woman, out of resistance, has not yet supplied me with the dates)" (1985, 183; see also 186, 191–92). All the while, by such reminders, Freud was in effect asking his friend to apologize—not to Eckstein, needless to say, but to him.

Emma Eckstein suffered from this doubly mistaken and careless treatment well into the summer of 1895, and she continued to see Freud. Obviously Max Schur grasped right away the significance of the episode to the "Irma" dream of the night of 23–24 July, since he first made these letters public in his paper on the specimen dream. But Schur treats the new material as "additional 'day residues'" contributing to the dream of Irma's injection, not as primary or as the main instigation of the dream. Since Freud himself, in his long analysis of the dream, had raised the possibility that the Irma figure might have stood for about nine different women, Schur is on home ground. "Could the hostility expressed against Irma also have been a displacement from Emma?" In real life, Schur seems confident, these were quite separate women. He also believes that he, and not Freud, has discovered their dream relation, or Freud would have included it in his analysis. "Knowing as we do that Freud wrote the final version of *The Interpretation of Dreams* in 1899, and knowing his superb honesty, we must assume that by that time any connection between the Irma dream and the Emma episode had been even more thoroughly repressed than before" (1966, 77, 73). Schur is good at making such connections but cannot seem to draw the most obvious conclusions. Freud's honesty apart, why wouldn't he have made the connection? Eckstein was very much on his mind that summer, as was Fliess. Besides, he covers himself against the charge of dishonesty by surrounding his published account of the dream with disclaimers, as we have seen: he never proposed to tell all. In Schur's subsequent biographical study, he states that "all these connections with the Emma episode indicate that the main wish behind Freud's Irma dream was not to exculpate *himself* but Fliess. It was a wish not to jeopardize his positive relationship with Fleiss" (1972, 87). This attempt to improve on Freud seems wrongheaded. The wish not to break with Fliess could hardly be anything but conscious, since it is so apparent in the letters, which deny that Fliess is to blame but pretty much remind him that he is—a not uncommon way of communicating with a friend. It would seem more likely that Freud ever since March had personally, consciously blamed Fliess for the disaster with Eckstein and his continuing problems with her, and that only after the dream of "Irma's injection" did he realize how much he was projecting his own responsibility, too, upon his colleague. The succeeding letters, after the dream, come down heavier on hysteria again—his own department. The dream, he thought, fulfilled a *wish* that "Irma's" problems were physical.

The Freudians who have continued in the wake of Schur to interpret the specimen dream also treat the knowledge of the Eckstein episode as supplementary. Alexander Grinstein (1968, 21–46) closely pursues a passing reference in Freud's interpretation to a novel by Fritz Reuter. Grinstein does not identify "Irma" outright, though almost the last thing he observes of the dream is that "actually" Freud had earlier thought about the bearing of wish fulfillment on dreams. The reference he cites happens to be another letter to Fliess that is largely about Eckstein. Didier Anzieu (1975, 131–55) uncovers much more sexual content but entertains so many fanciful notions about the dream that his interpretation becomes a kind of blur. He identifies "Irma" as Anna Lichtheim, née Hammerschlag, partly because of the similarity of names (if it comes to that, "Irma" is one letter closer to Emma). In the text of the draft project for the standard edition, Strachey carefully restores the holograph *A* where Freud refers to the dream (1950, 341n). At the time of editing, Strachey would have been familiar with Freud's cheerful confession, in a letter to Karl Abraham in 1908, that "the three women" of the dream, "Mathilde, Sophie and Anna, are my daughters' three godmothers" (1965, 20). Just recently, the specimen dream's affiliations with these three—the two young widows, Anna Lichtheim and Sophie Paneth, and Josef Breuer's wife, Mathilde—as well as Freud's possible feelings about his daughters have been meditated upon by Lisa Appignanesi and John Forrester (1992, 117–45). But as far as I know, Freudians have not attempted to start over again, with Schur's revelations and the new edition of the letters to Fliess in hand, in order to argue that "Irma," even if she appeared with the face and figure of Anna, may have been immediately recognizable to the dreamer as a displacement for Emma Eckstein. The chronology and the manifest dream both suggest that this was likely the case, and the tone of Freud's defiance as he withholds a portion of his interpretation in the dream book tends to confirm it. Those most interested in such matters tend to pile up interpretations, never to retract or complete them. A similar accretion of possibilities typifies Freud's own revisions of *The Interpretation of Dreams*, as well, and is undoubtedly more characteristic of hermeneutics than of science.

Two psychologists who care enough about psychoanalysis to compose an exhaustive survey of research attempting to confirm its theories manage to reach some interesting conclusions about the dream of Irma's injection. "A primary question that comes up in the course of such a scrutiny," according to Seymour Fisher and Roger P. Greenberg, "is the nature of the relationship that exists between what Freud called the manifest and latent dream contents. It is striking how willing most persons have been to accept Freud's view about the disparity between manifest and latent without any solid evidence except his strong but only impressionistically based assertions to this effect" (1985, 32). (In the next chapter I shall try to spell out precisely what is attractive in Freud's view and his immediate shift to the secret meaning of dreams.) Fisher and Greenberg devote a good deal of space to what conclusions might be deduced from a manifest dream alone, by applying the standard procedures of Rorschach analysis or thematic apperception tests; and to illustrate the point, they experiment with the text of Irma. A word count of Freud's dream suggests that it "is directly concerned with body themes largely involving illness and poor health," which they see as confirming some of Erikson's intuitions about the dream. Furthermore,

> if one analyzes the personal interactions in the explicit dream content, one finds that they can be differentiated into those involving unfriendly relationships between men and women and those depicting friendly ones between men. . . . The majority of male-female interactions involve hostility and threat. The male-male interactions are, contrastingly, friendly and mutually supportive. . . . To take this view would contradict Freud's own "latent" interpreta-

tion of the dream. He felt, on the basis of his private associations, that the dream contained considerable hostility toward significant male figures in his life. (1985, 72–73)

Again citing Erikson, these writers believe that Freud's relation
to Fliess was of concern to him in this dream. So their rather
mechanical means of interpretation yields a result consistent
with a conclusion that "Irma" primarily substituted for Emma
Eckstein.

Curiously, Freud actually wrote to Fliess on 24 July 1895—the
day, as he wrote five years later, that ought to be carved in marble—but without ever mentioning their patient or the dream or
his discovery. The letter is commonplace enough, and consists of
a paragraph chiding his friend for not writing and a pun about
living in heaven (Himmel was the name of the street in Bellevue): "Daimonie, why don't you write? . . . Are we friends only
in misfortune? Or do we also want to share the experiences of
calm times with each other?" and "Where will you spend the
month of August? We are living very contentedly in Himmel"
(1985, 134). Can Freud have mistaken the date? There is no sign
in the letter of his having dreamed of "Irma's injection" the
night before. Alas, no very firm conclusions can be drawn from
facts set forth by Freud, since he admittedly conceals facts, occasionally gets them wrong, and once or twice invents them. Some
of the barriers to comprehending his dream book, as we shall see,
are of a different order from the distinction he makes between
manifest and latent content, or a dream and its interpretation,
since they include a certain amount of wishful thinking.

CHAPTER TWO

"Dreams Really Have a Secret Meaning"

INSTEAD OF FIGHTING the dream book or being mystified by it, we can speculate usefully about what its author was hoping to do. What were the advantages of writing on dream interpretation? What is attractive about the theory chosen? Why should dreams have a secret meaning? What use is the search for motives? What is to be gained from basing narratives on the slightest evidence? There are no fixed answers to such questions, needless to say: one can merely interpret Freud's *Interpretation*. But wish fulfillment—that is, in story, not reality—is an excellent guide to understanding narrative, including the story of ambition and discovery in the dream book itself.

In contemplating the possible wish fulfillment of a book about the same, I do not mean to suggest its unconscious authorship, or anything like the force Freud assigned to repressed wishes. Still less do I mean to say that the author repressed whatever he did not write of his own motives (the way Max Schur believes he repressed what he knew of Emma Eckstein when he came to write of his dream). The wish fulfillment I have in mind is closer to that which Freud attributed to religion in *The Future of an Illusion* (1927). As long as certain prayers seem rational—that is, self-interested—the purpose of praying to Zeus can be appreciated by believers and unbelievers alike. The wishes that moved Freud to write *The Interpretation of Dreams* are all rational in this sense: he might not have been strictly aware of every wish I impute to his writing, but that is because he was bent on making the wishes come true, not because he repressed something. For Freud and those of us who read his work, Zeus was only a theory. Just so, as inheritors of the Enlightenment, we shall sooner or later submit Freud's theory to analysis. If this realization involves us in viewing his masterpiece as equivalent to some great nineteenth-century novel, that is because psychoanalysis undoubtedly derives some of its interest in the conditions of repression and scandal—and the revelation of secrets—from the novel.

The first surprising turn in Freud's argument antedates the book: at some point he decided to write about dreams, and this decision changed his life history. Besides being celebrated as a story of discovery and the beginning of a movement, *The Interpretation of Dreams* ought to be thought of as the master event in a career of writing. Not only is it Freud's longest single work but it was quickly followed by *The Psychopathology of Everyday Life* the next year, by *Jokes and Their Relation to the Unconscious* and *Three Essays on the Theory of Sexuality* in 1905, and thereafter by an extraordinary output of books, papers, and lectures until the end of his life. (Again the comparison with Darwin is striking. Freud was forty-three when the dream book appeared; Darwin, fifty when *On the Origin of Species* was published. Darwin, too, became a prolific writer in middle age, completing *The Descent of Man, and Selection in Relation to Sex* and eight other titles before his death.) In his preface to the second edition in 1909, Freud noted that the dream book had a personal significance that he realized only after it was written: "It was, I found, a portion of my own self-analysis, my reaction to my father's death—that is to say, to the most important event, the most poignant loss, of a man's life." But this loss and its indirect expression were fully consistent with the launching of his career as a writer. Jacob Freud died in the autumn of 1896, about a year after Freud had left off his draft project. This loss cleared a space for the son's career and helped determine its path: a good many of Freud's dreams—or, more accurately, his dream thoughts—bore on his relation to his father. Though a book may be aided or deterred by subjective states, however, the writing of it takes work. One must want to write a book, and being a writer entails deciding what to write about. Freud's decision to write the book presumably preceded his father's death, since he later claimed it "was finished in all essentials at the beginning of 1896 but was not written out until the summer of 1899" (1914, 22). That he was writing "fluently" by February 1898 we can be sure from his letters to Fliess (1985, 298). At some point after abandoning the draft project, Freud determined to write on dreams.

The decision was an important one in middle life. It meant turning for a while at least from physiology to psychology, a desertion of the usual precincts of medical science; and it posed the opportunity of extending what he had observed of unconscious motives in neuropaths to normal people. There were undoubtedly some practical advantages in studying dreams. In the field of neurology he was up against limitations of experiment and instrumentation that exist to this day. The only data he could realistically hope to amass were the highly subjective gleanings of clinical work, on a limited number of patients. The data base of dreams was virtually unlimited; dreams occur every night, apparently to everyone. That this new base was unreliable, fragmented, and at least as subjective as clinical observations was not so disadvantageous to a writer of a book as it might seem to a research scientist. The quirky, unstable evidence of dreams obviously provided, and provides, the dream interpreter and theorist great leeway. Freud could only work with the manifest content of a dream "as it is presented in our memory" (277). No one could ask for more, but he would give more: the latent content. These inherent advantages of his new project, the writing of a book on dreams, can be shared by its readers, even as they partake in the sophisticated pleasures of detecting wish fulfillment. Few readers get as far as Freud's specimen dream without wondering what the relation of the script can be to the dream he experienced. But we appreciate from our own experience how tenuous such memories must be and are poised to learn what the writer will do with them. Above all, this sort of evidence is familiar, for if lay persons have little or no clinical experience, all have had dreams.

That "*a dream is the fulfilment of a wish*" (121) seems to have been Freud's first conclusion, the immediate inspiration of his analysis of Irma's injection. If one steps well back from the argument, one can speculate how this thesis itself might be motivated, or what pleasure might be anticipated from it, by comparing it to the theses of other conceivable dream books. A thesis that dreams were messages from the gods might be pleasurable, for example, but the pleasure would depend largely on the gods. Freud's thesis preserves the independence of the dreamer, in an era when the gods are dead and dreamers enlightened. One should also compare an alternative thesis that dogs Freud's footsteps everywhere in this book: namely, that a dream could be the expression of anxiety. This is a distinctly less pleasurable thesis than wish fulfillment; the emotions that this thesis would point to, especially if the dreamer were paterfamilias, are discomforting. At one point, introducing his childhood dream of the bird-beaked figures, Freud remarks that "it is dozens of years since I myself had a true anxiety-dream" (583). Thus he admits that anxiety can be experienced in the manifest dream but perhaps more importantly exempts himself from the common experience of most adults: he no longer fears anything, even in his sleep. I am directing these conjectures not ad hominem but as a way of getting at the pleasures of the thesis. *The Interpretation of Dreams* is a masterly performance because, among other reasons, it invites its readers in so many ways to identify with Freud. The pleasure of the text is pleasure shared. Think of the difference it would make if we were asked to share the writer's anxieties rather than his wishes. We could manage it, obviously, but it would be like a very long trip with Dostoevsky's underground man or Kafka's Joseph K.

The reader's point of entry to the dream book is at the side door, with the doctor and not with the sick. That Freud is also prepared to investigate his own dreams only enhances the reader's identification. Shared pleasures almost certainly influence the selection of dreams and interpretations, and most likely the thesis that a dream is the fulfillment of a wish. Both the dreams in Freud's book and his interpretations are all very temperate, surely. Nightmares are not included, and deep anger never seems to underlie the dreams any more than fear does. Many frustrations and petty annoyances come into play, but such emotions are in the last analysis more respectable than anger or fear. Admittedly, discovering wish fulfillment in so many different scenarios affords a rather subtle pleasure, one that is not immediately self-evident. Indulging in wish fulfillment is generally regarded as a failing, though not a serious failing (it might be serious, of course, in a political program or scientific theory, or lead to trouble in personal relations). Freud knows that; but significantly he is not at all contemptuous of wish fulfillment. If he were contemptuous of it, the charm of the book would vanish, and I doubt he would have completed it. He treats wish fulfillment as a foible; he indulges this very common self-indulgence. What, then, is the pleasure attached to the thesis, or to the repeated discovery of wish fulfillment in the famous book? The best general word for it may be *knowingness*. It is a kind of knowingness that Freud makes it possible for his readers to share.

The passion of knowingness is to be in on secrets; and its pinnacle, the power to reveal them. The numerous interpretations collected in the dream book, short and long, satisfy that passion, while identification with its author can only add to the pleasure. Even the discovery of the theory Freud will memorialize as the revelation of a secret: "Here, on July 24, 1895, / the secret of the dream / revealed itself to Dr. Sigm. Freud" (1985, 417). As I have suggested, the critical turn in the book is from wish fulfillment per se to the revised formula at the end of the chapter on distortion: "*a dream is a (disguised) fulfilment of a (suppressed or repressed) wish*" (160). The satisfaction inherent in this turn—its emotional return, so to speak—is already apparent at the end of the previous chapter when the writer casually anticipates "our theory of the hidden meaning of dreams" (132) or when he cautions, "The fact that dreams really have a secret meaning which represents the fulfilment of a wish must be proved afresh in each particular case by analysis" (146). The scrupulousness about proof shrinks in comparison to the promise that each dream has a secret meaning; or rather, both propositions in this last sentence are rhetorically effective, and their bold combination of promise and scruple is greater than the sum of its parts. Rhetoric or substance, the value of secrets for Freud's theory can hardly be exaggerated. Unless the meaning of dreams were hidden, there would be nothing for Freud to do or for his readers to enjoy—hence he very quickly gets over those childish dreams of undisguised wishes and saves for later the deep secrets of childhood. The claim to reveal secrets pervades the book, which cheerfully confuses the secrets of nature, things unknown to anyone before now, with the secrets of personality, things concealed from others and more especially from oneself. All this interest is nicely summed up when Freud introduces the theoretical matter of his last chapter: "Hitherto we have been principally concerned with the secret meaning of dreams and the method of discovering it and with the means employed by the dream-work for concealing it" (Wir haben uns bisher vorwiegend darum gekümmert, worin der geheime Sinn der Träume besteht, auf welchem Weg derselbe gefunden wird, und welcher Mittel sich die Traumarbeit bedient hat, ihn zu verbergen; 510). Note that the language by now assumes that concealment is natural, while Freud also regularly assigns to it specific human motives. The claim to disclose secrets is a powerful one.

A frequent complaint used to be that the only secrets that interested Freud were sexual. The dreams collected in the book belie this, and the author prefers in the end to "leave it an open question whether these sexual and infantile factors are equally required in the theory of dreams" (606). Far more telling is a parenthetical remark by Wittgenstein, just where he complains of the "muddled" assertion that all dreams are wish fulfillments: "Freud very commonly gives what we might call a sexual interpretation. But it is interesting that among all the reports of dreams which he gives, there is not a single example of a straightforward sexual dream. Yet these are common as rain" (1966, 47). Between the missing sexual dreams and the frequent sexual interpretations lies the Freudian program, with its nearly exclusive interest in disguised or secret wishes. Although it may be true that dreams of having sex are as common as rain, the wish therein represented is in no need of interpretation. Freud in fact does interpret some dreams as signaling a straightforward desire for sex, but these are almost all women's dreams. Often dreams of falling are "characterized by anxiety," for example, but women "almost always accept the symbolic use of falling as a way of describing a surrender to an erotic temptation" (394–95). An exception to the rule that dreams attributed to men are not usually about sex is provided by one man who dreams of enjoying his wife "*a tergo*" (397), but that secret wish merely calls attention to the absence of straightforward sexual dreams among the men. This gender distinction at first seems contrary to the social code, which allows that men want sex more than women do, but it is easy to see how this disparity comes about. Freud does not necessarily believe that women want it more (though there is a little of that); rather, women's desires of this kind are a well-kept secret, even from themselves. A book about the hidden meaning of dreams published in 1900 was bound to privilege women's dreams of sex over men's. In this, as in most respects, the dream book and the social code are quite in harmony.

Concealment provides the drama of the dream book. In every way he can, Freud urges that the wish behind a dream—a target of analysis—is not merely unknown, like the object of some less touchy scientific inquiry, but fearfully defended. The secret meaning of dreams means to be secret: this active component in the unknown materials heightens the pleasure of finding them out. One meets with it everywhere in *The Interpretation of Dreams*—in the metaphor of censorship, certainly, with its imputation of a specific directive to conceal something, but also in the many casual metaphors of something forbidden or proscribed. The method of interpreting dreams is supposedly scientific, but the *object* of its study regularly adopts all the deliberate guises of a conspiracy. Freud's favorite term is *resistance* (Widerstand), which, far from being a metaphor from electrical engineering, is repeatedly used to humanize the contest of doctor and patient, or the analyst and the dream. Even—or especially—if a dream is forgotten, it has a secret meaning: "Psycho-analysis is justly suspicious. One of its rules is that *whatever interrupts the progress of analytic work is a resistance*" (517). Analysis thus temporarily becomes adversarial, while at the same time mystery is added to the drama. Such "rules" (Regeln), instead of constraining analysis, typically give it more license. In a rare discussion of appropriate technique, Freud recommends fixing immediately on any portion of a dream that has been forgotten and then remembered:

> It not infrequently happens that in the middle of the work of interpretation an omitted portion of the dream comes to light and is described as forgotten till that moment. Now a part of a dream that has been rescued from oblivion in this way is invariably the most important part; it always lies on the shortest road to the dream's solution and has for that reason been exposed to resistance more than any other part. (518–19)

To argue so assumes a previous though unconscious intent on the dreamer's part to conceal something; and of course if concealment itself is significant, it can only be as a sign of guilt. The reasoning behind the suggested technique is of the prosecutorial kind that is not likely to reveal anything but guilt and that is so apparent in Freud's own case study of Dora in this same period (1905b).

Freud himself was not slow to notice the similarity between his method of probing for secret meanings and the practice of criminal investigation. Though he wrote little about the parallel, he did contribute a guest lecture to a law seminar in Vienna, on the "objective signs" of guilt sought for by the two professions.

"The task of the therapist," he suggested, "is the same as that of the examining magistrate. We have to uncover the hidden psychical material; and in order to do this we have invented a number of detective devices, some of which it seems that you gentlemen of the law are now about to copy from us" (1906, 103, 108). In these sober comments to law students, he depended mainly on the idea of resistance and touched on some of the ways— including special attention to freshly recalled portions of dreams—for penetrating resistance and thus turning it into a form of self-betrayal. The talk also included some sensible cautions on the differences between psychotherapy and criminal investigation. Freud, of course, more than hints that psychoanalysis has precedence as far as this new interest in criminology is concerned: he wishes to bring to the lawyers' notice "that an exactly similar method of disclosing psychical material which is buried away or kept secret has been practised for more than a decade" in his own field (1906, 107). Actually, criminal psychology was already a popular subject. In his lecture, Freud speaks respectfully, if tangentially, of Hans Gross, then professor of criminal law at Prague and later at Graz, and formerly an examining magistrate. Without appealing to any systematic depth psychology, Gross boldly asserted that secret motives could be known by their disguises. His whole book contended that criminals frequently gave themselves away. A long section on women (compare Freud's view of Dora) suggests that they may be easier to see through than men, because of their inveterate sexual needs. Whether she is a defendant or a witness, Gross explains, "We must discover whether a woman is morally pure or sensual, etc." But we need not confine ourselves to direct evidence. Three indirect signs of a woman's concealed sexual passion that Gross discusses at length are any sudden piety, marked ennui, or increased conceitedness (1905, 322–32). His *Criminal Psychology* was first published in 1897 and achieved four editions by 1910. Henri F. Ellenberger (1970, 495) puts forward an earlier work by Gross as a possible inspiration for the Freudian theory of slips of the tongue. My own persuasion (1992a, 147–50) is that Freud's customary practice—his search for indirect evidence of unconscious wishes and his "justly suspicious" approach to resistance—would not have arisen without a century or more of heightened interest in criminal prosecution.

A striking illustration of Freud's attitude toward the concealment of guilt, and of silence or denial as evidence of guilt, can be found in his subsequent *Introductory Lectures*. Freud there defends himself against the argument that he characteristically accepts from patients what he wants to hear but refuses to accept their denials. Specifically, he ironically imagines a little interchange with his listeners on the subject of parapraxes, then draws a parallel to a routine that he regards as both commonplace and justified in a criminal examination.

> "So that's your technique," I hear you say. "When a person who has made a slip of the tongue says something about it that suits you, you pronounce him to be the final decisive authority on the subject. 'He says so himself!' But when what he says doesn't suit your book, then all at once you say he's of no importance—there's no need to believe him."
>
> That is quite true. But I can put a similar case to you in which the same monstrous event occurs. When someone charged with an offence confesses his deed to the judge, the judge believes his confession; but if he denies it, the judge does not believe him. If it were otherwise, there would be no administration of justice, and in spite of occasional errors we must allow that the system works.

That the system works argues that it ought to be an excellent model for psychoanalysis. At the same time, Freud's sarcasm—his allusion to this "monstrous" practice—shows how readily he identifies with the prosecutorial side of the administration of justice. When his imagined listener rejoins, also sarcastically, "Are you a judge then?" he answers stiffly, "Perhaps we need not reject the comparison" (1916–17, 50). But Freud as judge anticipates the outcome. The logic behind the supposed judicial inference of guilt depends mainly on a prior persuasion, or separate grounds for believing, that the subject is guilty. Only so can a denial be dismissed, and never out of hand.

It is unlikely that Freud needed to spend much time thinking directly about criminal cases. In the second half of the nineteenth century, novels were the main depository of wisdom about criminal guilt and its detection. Jones informs us, for example (1953, 1.174), that Freud knew and admired both *Middlemarch* and *Daniel Deronda*, novels by George Eliot not less concerned with the effects of concealment of guilty thoughts and secrets of the past than psychoanalysis would become a generation later. Most obviously, detective novels took for granted a social dynamics of scandal and its suppression and routinely enacted a search for hidden evidence of complicity in a suspect's history. Such novels provided a paradigm, social and in some instances psychological, for Freud's discoveries and his occasional prosecutorial tone. Secrets of the past became fashionable material for realistic as well as sensational literature. (Two quite serious and pervasive ideas, one scientific and the other religious, underwrote both novels and psychoanalysis: the new historical sciences of geology and evolutionary biology reinforced the idea of a universe totally caused, down to the minutest effects, in space and time; religion—or, if not, superstition—cherished the notion that sooner or later all secrets, but especially crimes, would become known.) The great detective was one who, like Dickens's Inspector Bucket, could recognize the telltale signs of whatever had been concealed. If the detective was on the alert, the slightest evidence would avail, since everything signified. Wilkie Collins's Sergeant Cuff, in all his practice, "never met such a thing as a trifle yet" (1868, 136). A very similar view of evidence was expressed by Freud, who writes in the dream book, "Examples could be found in every analysis to show that precisely the most trivial elements of a dream are indispensable to its interpretation and that the work in hand is held up if attention is not paid to these elements until too late" (513). Even so, Inspector Bucket overlooked one detail, set off on a wrong course, and caught up with Lady Dedlock too late; Sergeant Cuff traced Rosanna Spearman's involvement accurately but was blind to her romantic motive, and it took the medical assistant Ezra Jennings to show that the hero himself had taken the diamond. Both famous detectives followed one sure rule, however: in calculating a woman's moves, they counted on her resistance and penchant for secrecy. Yet the single scrap of evidence that can only be accounted for by a specific crime or wish is surrounded by a thousand other signs that may also be evidence, in this crowded universe of interconnected matter. The exaltation of "a trifle" or of "the most trivial elements of a dream" is finally a tribute to the skills of the detective or analyst who can recognize their unique significance. The slighter the evidence, to be sure, the grander the feat of detection or analysis and the happier the reader.

Besides an assumption of guilt, which has important secular as well as religious roots in the nineteenth century, the explanatory power of a motive, around which a plausible narrative can be reconstructed, would seem essential for the development of psychoanalysis. It is unlikely that fixing a motive (Motiv) could be such an important step in Freud's dream interpretation if it were not for the myriads of shrewd attempts in the nineteenth century—whether legal or popular—to fix individual responsibility independently of religion. Supplying a motive as the key to the dream thoughts is something that Freud goes about automatically. It is his way of listening to patients, his way of catching himself—his mode of explanation and interpretation. Freud's method might splendidly have served as the inspiration of his own favorite Dickens novel, *David Copperfield*, which was originally to have been about a man who habitually thought, "Well, now, yes—no doubt that was a fine thing to do! But now, stop a minute, let us see— *What's his motive?*" (see the respective biographies by Ernest Jones, 1953, 1.174; and John Forster, 1872–74, 2.77). Characteristically, even as Freud interprets his dream of Irma's injection, he asks in respect to himself and Dr. M., "But what could be my motive for treating this friend of mine so badly?" (115). Such questioning leads to the complex motive that is the explanation of the dream and the basis of the theory:

> The conclusion of the dream, that is to say, was that I was not responsible for the persistence of Irma's pains, but that Otto was. Otto had in fact annoyed me by his remarks about Irma's incomplete cure, and the dream gave me my revenge by throwing the reproach back onto him. The dream acquitted me of the responsibility for Irma's condition by showing that it was due to other factors—it produced a whole series of reasons. The dream represented a particular state of affairs as I should have wished it to be. *Thus its content was the fulfilment of a wish and its motive was a wish.* (118–19)

And thus Freud's subsequent shift to disguised wishes is also a shift to secret motives that—as in the Dickensian paradigm of suspicion—may give the lie to appearances.

That motive is the sine qua non of such interpretations is clear from the terms in which Freud dismisses somatic explanations, which are "incapable of producing any motive governing the relation between an external stimulus and the dream-idea." External stimuli might have wholly other effects besides the dream, such as waking the sleeper entirely; therefore "the motive for dreaming" lies elsewhere (223–24). Common usage supports his equation of wish and motive, since neither is simply desire but compounded of desire and belief in some sequence of events that will satisfy the desire. Richard Wollheim is helpful here, since without noting Freud's persistent appeal to motive, he generalizes in similar terms about explanation in all the psychological works. Wollheim suggests that Freud deepens, elaborates, and varies—most notably by introducing unconscious desires and beliefs—this mode of explanation. "It is further to be noted that not only does the citation of a desire and a belief tied in this way to an action explain that action, it also establishes it as an action. An action, it has been said, is something that we do about which it is possible to ask in the appropriate sense *why* we did it." As to why Freud adopted this scheme, Wollheim seems to feel there can be "no alternative" (1990, xxx–xxxii). This mode of explanation, however, also has a history in modern times. In Freud's case it may owe something to Schopenhauer's distinction between the will and motivation (1844, 1.294), but much more broadly it reflects the need to define responsibility apart from obedience to any external influence, divine or human. Wollheim's observation that this conjunction of desire and belief "establishes . . . an action" is the key: in the absence of externally imposed law, it becomes in theory a nice problem to exact responsibility on the part of individuals, which is what is meant here by belief that establishes an action—something that rises to an action, for which one is accountable. Not surprisingly, the problem is most clearly addressed by the criminal code—that is, post-Enlightenment criminal codes, which forego divine authority and in theory create institutional authority only by a social contract.

The importance of motives in *The Interpretation of Dreams* is most evident when one recalls the degree to which belief, in Wollheim's sense, inhabits the Freudian unconscious. It may seem counterintuitive that beliefs, which can be expressed as linguistic propositions about a limitless number of real conditions, should be unconscious even in the literal sense of unknown, but such is one of the earliest tenets of Freud's thinking. Josef Breuer wrestled with it in the theoretical part of their joint enterprise, *Studies on Hysteria* (1895, 222–25). Twenty years later Freud still emphasized, contrary to what one might suppose, that instincts as such could be neither conscious nor unconscious. Only ideas, representations of some kind, move between these two states.

> Even in the unconscious . . . an instinct cannot be represented other than by an idea. If the instinct did not attach itself to an idea or manifest itself as an affective state, we could know nothing about it. When we nevertheless speak of an unconscious impulse or of a repressed instinctual im-

pulse, the looseness of phraseology is a harmless one. We can only mean an instinctual impulse the ideational representative of which is unconscious, for nothing else comes into consideration. (1915, 177)

This principle, which seemed strained when Breuer introduced it, facilitated the explanatory process of psychoanalysis. An instinct by itself might be taken for granted, but unless it were accompanied by some idea—of how it might be served or frustrated, say—no imaginary event would occur such as literally could be narrated. Something like this is close to what we mean by establishing a motive, and the instinct attached to an idea—the desire-belief model again—allows for this.

Freud's reliance on free association brings additional weight to bear on some underlying idea, if a convincing explanation or coherent narrative is to be achieved. A protracted defense of the method in chapter 7 refers no longer to motives but to "purposive ideas" (Zielvorstellungen), which similarly compound desire and belief. Here, as "no connection was too loose, no joke too bad, to serve as a bridge from one thought to another," Freud seems acutely aware of the tendency of dream analysis itself to disintegrate. But "the real reason for the prevalence of superficial associations is not the abandonment of purposive ideas but the pressure of the censorship." Apparently, because the purposive ideas (wishes, motives) are forbidden or distressing, they emerge only in a disguised, often trivialized form. The appeal at this point is frankly from dream interpretation to clinical work:

> In the psycho-analysis of neuroses the fullest use is made of these two theorems—that, when conscious purposive ideas are abandoned, concealed purposive ideas assume control over the current of ideas, and that superficial associations are only substitutes by displacement for suppressed deeper ones. Indeed, these theorems have become basic pillars of psycho-analytic technique. When I instruct a patient to abandon reflection of any kind and to tell me whatever

comes into his head, I am relying firmly on the presumption that he will not be able to abandon the purposive ideas
inherent in the treatment and I feel justified in inferring
that what seem to be the most innocent and arbitrary
things which he tells me are in fact related to his illness.
There is another purposive idea of which the patient has no
suspicion—one relating to myself. (530–32)

The passage becomes a compendium of the method, fraught
with the usual question of who possesses the ideas "inherent in
the treatment." What I wish to stress is the way that the whole
discussion, even the reference to what came to be called the
transference, leans on the notion of purposive idea, something
that will meaningfully pull together the relevant details.

Also at the beginning of chapter 7, Freud anticipates and skill-
fully articulates the very arguments that Wittgenstein and others
would bring against the logic of free association. "There is nothing wonderful in the fact that a single element of the dream
should lead us *somewhere*," he imagines his critics saying; "every
idea can be associated with *something*. What *is* remarkable is that
such an aimless and arbitrary train of thought should happen to
bring us to the dream-thoughts." In this summary of the opposing view, he virtually concedes the likelihood that some linking
idea or other is bound to fit with "a chain of associations." Thus
it appears that "the whole thing is completely arbitrary; we are
merely exploiting chance connections in a manner which gives
an effect of ingenuity. In this way anyone who cares to take such
useless pains can worry out any interpretation he pleases from
any dream" (527). Freud's repeated use of the metaphor of a
chain in this construction of the opposing argument is reminiscent of "the chain of circumstances" popularized in criminal tri-
als. His reply to the objections, in fact, echoes a classic argument
for the trustworthiness of circumstantial evidence. "We might
defend ourselves," he writes,

> by appealing to the impression made by our interpreta-
> tions, to the surprising connections with other elements of

the dream which emerge in the course of our pursuing a single one of its ideas, and to the improbability that anything which gives such an exhaustive account of the dream could have been arrived at except by following up psychical connections which had already been laid down. (527–28)

As far as I can see, the first two defenses in no way get past the objections that Freud himself has spelled out; the third, however, adopts an argument that had become fairly commonplace by the eighteenth century (1992a, 28–42). The argument is that a truly "exhaustive account" of circumstances must not only cohere within itself but correspond with reality at so many different points that it could not be invented.

In this case Freud rests his case on probability—which his language seldom admits. Ian Hacking (1975, 1990) and others have reminded us that the whole modern argument for preferring facts to testimony in science assumes probabilistic thinking; I happen to think that Freud's position is even closer to the parallel course of criminal prosecution and forensic argument in the same period (and much more work should certainly be done on this subject). Though at first glance psychoanalysis seems to rely exclusively on testimony, because the patient speaks at length to the analyst, in fact it relies on a reconstituted narrative of circumstantial evidence. Again, the forensic parallel that Freud constructs in the *Introductory Lectures* recalls the social history of this way of thinking:

I suggest that you shall grant me that there can be no doubt of a parapraxis having a sense if the subject himself admits it. *I* will admit in return that we cannot arrive at a direct proof of the suspected sense if the subject refuses us information, and equally, of course, if he is not at hand to give us the information. Then, as in the case of the administration of justice, we are obliged to turn to circumstantial evidence, which may make a decision more probable in some instances and less so in others. In the law courts it may be necessary for practical purposes to find a defendant guilty

on circumstantial evidence. We are under no such necessity; but neither are we obliged to disregard the circumstantial evidence. (1916–17, 50–51)

Note how, even as Freud bargains for the use of indirect proof
or circumstantial evidence in psychoanalysis, he first places confession ("if the subject himself admits it") safely beyond dispute
once more. He has not really surrendered the position that confessions of guilt are valid and protestations of innocence are not,
or mentioned the possibility that patients and prisoners sometimes admit to what they are asked regardless.

In general, psychoanalysis distrusts testimony ("psycho-analysis is justly suspicious"). The patient's resistance must be taken
into account, his or her free associations have to be reorganized
as a coherent narrative, and that which supplies coherence is a
purposive idea or motive. The evasions, distortions, and refusals
of testimony provide only indirect evidence, which must be ar-
ranged in such a way as to get beyond the purely subjective opinion of the witness. Similarly, a manifest dream must be retold as
the latent dream. The privilege of reordering a dream "cut up
into pieces" (103) is one that Freud has claimed in *The Interpretation of Dreams* from the start; the purpose of the dissection is to
replace the dream with an objective narrative. He characterizes
the reordering of materials as scientific, for "the work of interpretation is not brought to bear on the dream as a whole but on
each portion of the dream's content independently, as though
the dream were a geological conglomerate in which each fragment of rock required a separate assessment" (99); and those independent assessments serve as a check to error in the reconstitu-
tion of the whole, as insurance against invention. Freud is aware
that this procedure is not limited to science: he associates it, for
example, with one of the traditional ways of interpreting dreams,
which he calls decoding; and when he addresses the law seminar
in Vienna, he begins by conceding "the untrustworthiness of
statements made by witnesses" and contrasts the modern aim of
establishing "guilt or innocence by objective signs" (1906, 103).

The aim throughout is objectivity; and even if we conclude that Freud's theory is only a speculation, it pursues a characteristically modern quest for objectivity.

By the late nineteenth century, objectivity could be enhanced by limiting the motives available to interpretation, whether of behavior or of a dream. That people sometimes—and saints frequently—perform charitable acts of self-sacrifice, are kind to one another, or are generous to their enemies cannot really be explained objectively anymore, except perhaps as sublimation of a sexual instinct or collective self-interest. If a dream, therefore, seems to represent an altruistic impulse, or the fulfillment of a charitable wish, its latent meaning must be otherwise. The possibility that we sometimes wish good for others occurs to Freud, for several times he remarks upon it as incompatible with the dream thoughts. Dreams "are completely egoistic: the beloved ego appears in all of them, even though it may be disguised. The wishes that are fulfilled in them are invariably the ego's wishes, and if a dream seems to have been provoked by an altruistic interest, we are only being deceived by appearances" (267). Hence dreams can never be dreamed for others, as a social worker might dream; properly interpreted, they tell the objective truth about the subject. "Whenever my own ego does not appear in the content of the dream, but only some extraneous person, I may safely assume that my own ego lies concealed, by identification, behind this other person" (322–23). Of course, this position may be forced upon Freud, since he is fixed so determinedly on the secret meaning. What would be the point of concealing, even from oneself, a kindly wish?

The focus on concealment in the book effectively narrows the available dream thoughts not merely to the ego's wishes but to those egoistic wishes that would be socially disapproved—antisocial wishes, as it were. Freud's masterpiece does not address rough exteriors with hearts of gold but only the opposite hypocrisies. With so much stripping away of appearances, in fact, the book becomes a satire of sorts. At the same time, its standard of objectivity is very much of the nineteenth century. Darwin

believed that individuals and whole species depended upon one
another, but only for selfish ends: "I do not believe that any
animal in the world performs an action for the exclusive good of
another of a distinct species." In a universe without purpose, the
mode of explanation rules out altruism: "What natural selection
cannot do, is to modify the structure of one species, without
giving it any advantage, for the good of another species; and
though statements to this effect may be found in works of natural history, I cannot find one case which will bear investigation"
(1859, 211, 87).

For the authors of such ambitious works as *On the Origin of
Species* and *The Interpretation of Dreams*, an obvious wish to be
fulfilled was precedence in the field and the achievement of lasting fame. The claim of precedence is transparent in both these
texts and can be taken for granted in most scientific publication
of the modern era. Once the program of psychoanalysis had
made some headway, Freud quickly sought to consolidate his
gain by institutionalizing the practice—providing leadership, re-
fining the doctrine, guarding against secession. These moves,
many of which were detailed by Freud himself (1914), have admittedly found their critics, including most recently Phyllis
Grosskurth (1991). By the second edition of *The Interpretation of
Dreams*, however, the author was prepared to cast himself as the
martyr to his own cause, as the 1909 preface and a special post-
script to the first chapter attest. Of course, the scientific precedence sought for is a form of property, a right to fame that out-
lasts life itself. For this purpose, Freud's masterpiece, insofar as
it began psychoanalysis, succeeded beyond his dreams, and to a
certain extent this general wish was fulfilled by the argument of
the book itself.

Fame was the design of those same insistent universal propositions in the dream book that have drawn protests. At least two
professionals who knew Freud believed that a penchant for over-
statement was part of his character. According to Jones, "When
he got hold of a simple but significant fact he would feel, and
know, that it was an example of something general or universal,

and the idea of collecting statistics on the matter was quite alien to him" (1953, 1.97). Breuer similarly characterized the man in his lifetime: "Freud is a man given to absolute and exclusive formulations: this is a psychical need which, in my opinion, leads to excessive generalization" (in a letter quoted by Paul F. Cranefield, 1958, 320). Confirmation, if any is needed, may be given by Freud himself, in the strange logic by which he defends the claims of his dream book in his analysis of Dora (1905b, 68). But it is not necessary to establish a character trait in order to understand the attractiveness, for writer and readers alike, of those excessive generalizations. In the dream book Freud typically ventures a universal proposition and then, instead of modifying it, caps it with another. The purpose—if one may judge by the effect—is to enter such sweeping claims that one cannot help taking notice. A theory of dreams must be commensurate with the laws governing the circulation of the planets or with the theory of natural selection.

To reveal the secret meaning of dreams, however, suggests powers more than scientific. Freud makes us aware that he takes up where the ancient interpretation of dreams left off. His scientism promises to correct the supernaturalism and superstition of older ways, without quite subduing the thrill of putting us in touch with secrets. As I have argued, secrets presuppose concealment, and therefore resistance. A dramatic struggle for the truth commences. Moreover, the secrets are the disagreeable truths that are hard to face. So the pleasure of this discovery is superior, a triumph over appearances and specifically over the many pettier pleasures of lesser wish fulfillments. Knowingness is a form of self-gratification inhabiting the least intellectual pretension, and Freud's is far from least. Guilt, as the favorite object of knowingness, may be as enjoyable as a public accusation of wrongdoing would be terrible. But the strongest appeal of Freud's masterpiece may be, in a special sense, evidentiary. It is his subjecting of the single clue, the slightest connection, the nearly forgotten past—any minimal piece of evidence, in short—to confident analysis that is bound to impress his readers.

Interpretation becomes a *feat* of inference and inspired guess-work, a little like a magical performance. Just so, the ingenuity of the interpretations in the dream book will culminate in admiration for the performer.

Even as he claims precedence and achieves mastery, Freud regularly depreciates his own genius by stressing the hard work of analysis instead. As David Riesman has argued (1950), the work ethic of psychoanalysis is scarcely less pronounced than that of Puritanism. "No one should expect that an interpretation of his dreams will fall into his lap like manna from the skies," Freud warns. One who seeks to interpret dreams "must familiarize himself with the expectations raised in the present volume and . . . must bear in mind Claude Bernard's advice to experimenters in a physiological laboratory: 'travailler comme une bête'—he must work, that is, with as much persistence as an animal and with as much disregard for the result" (522–23). The dedication to work recalls the sentiments of that other anti-Puritan genius, Dickens, who offered comparable advice to would-be writers and whose Copperfield modestly attributed his fame to hard work rather than any special talent for writing that he might possess.

CHAPTER THREE

"So Far as I Knew, I Was Not an Ambitious Man"

THOUGH TRAINED psychoanalysts often say that chapter 7 delivers the important lessons of *The Interpretation of Dreams*, far more readers respond positively to Freud's attractive self-presentation. One of the ways Freud most pleases—however testily he complained about supposed indiscretions—is by confession of his own dreams. These dreams, with the background and analysis he provides, are absorbing in themselves and partially linked up as autobiography. Only the naive or the doctrinaire, however, can suppose that the success of this broken narrative is due to its honesty per se. Confessions are by definition formally honest, since unless they reveal what a person can be expected to conceal they are not called confessions. But Freud aims to please, and that aim at once serves as a check on revelation and takes pleasure in the act. As he frequently acknowledges, he holds back some things of which, presumably, we would not approve. More than that, for every discreditable wish or story Freud tells about himself—usually not very heinous and never harmful to anyone—he regularly gains credit for self-knowledge and good manners.

It is easy to see how this credit builds, from the specimen dream to the very last anecdote in the book. The dream of Irma's injection is about responsibility, a fulfillment of the wish "that I was not responsible for the persistence of Irma's pains, but that Otto was" (118). Freud's meticulous interpretation, besides testifying to his truth seeking, implies throughout that he is a fully responsible professional person. To chide himself for wanting to evade responsibility is to assure himself, and his readers, that he knows well how to accept responsibility and that to evade it is wrong. The professional and social signals being sent are all very positive. He need not and does not admit that he is responsible for injuring anyone except in a dream, the interpretation of which proves his right-mindedness. The interpretation does not constitute a full confession, as we may say, but rather the confession of a wish, which has instigated a dream and not an act. Freud admits to having "revenged" himself (115), to having accused Otto of "thoughtlessness" (117), and to having pleaded his own "exculpation" (120) in his dream. Thus we know that he knows such thinking is in bad form and could only be expressed in his sleep. And these gestures and reassurances will be repeated, as in his dream of the botanical monograph: "Once again the dream, like the one we first analysed—the dream of Irma's injection—turns out to have been in the nature of a self-justification, a plea on behalf of my own rights" (173). There can be nothing intrinsically wrong with self-justification or asserting one's own rights; Freud is merely confiding that either may be regarded as bad manners. The claim that such is the secret meaning of his dream implies that his ordinary waking manners are excellent. Freudians who cherish the confessional prerogatives of the dream book as evidence of the author's fearless honesty forget, it seems to me, both what is being confessed and the print medium. The author of a book is not communicating with a priest or a trusted associate but to the public, to as many readers as he can find. Once again the principle of wish fulfillment is a better guide to Freud's purposes and extraordinary success. As self-presentation, one may judge, *The Interpretation of Dreams* portrays its author as he wanted to appear.

Repeatedly the book displays good humor and good manners. In the midst of interpreting the botanical monograph Freud quips, "There needs no ghost, my lord, come from the grave /To tell us this" (175). The author is gracious to joke about "my friend Otto, whose fate it seems to be to be ill-treated in my dreams" (271). It is an uncommonly attractive self that presents itself so. At the same time, Freud is never the man seriously to blame himself, as he sometimes implies he is about to do. Though he does not seriously blame Otto, either, deep down he is even more sure of his own blamelessness, with which readers may also identify. How he protects himself may be illustrated by the "Errors" chapter in *The Psychopathology of Everyday Life*, where Freud examines a number of factual mistakes in the dream book. He mistook Marburg for Marbach, Schiller's birthplace; Hasdrupal for Hamilcar, Hannibal's father; Zeus for Kronos, the emasculating son; and therein he would seem to be remiss. "I was responsible for a number of falsifications which I was astonished to discover after the book was published." Again responsibility is the theme, and again confession has its pleasures. "After closer examination I found that [the falsifications] did not owe their origin to my ignorance, but are traceable to errors of memory which analysis is able to explain" (1901, 217). Some writers would be ashamed of such mistakes, and most would cut their losses and shrug them off. Whatever one thinks of Freud's explanations, two of which are famous, he manages to make the errors in his previous book redound to his credit. He erred as he wished, to be sure; his subsequent explanations then satisfy his wish not to be ignorant and—as any writer can attest—not to be shown up.

In *The Psychopathology of Everyday Life* Freud makes a point of having known the right facts all along. His followers have usually been persuaded that the significance of his "Errors" chapter lies in the analyses, which generate narratives similar to the latent content of dreams. But even if one is grateful for these stories, a handful of distortions of known matters of fact (no longer to be thought of as mindless mistakes) argue that there must be a great many facts in the dream book that have been distorted by other purposive ideas. Significantly, Freud copes with three errors visible to anyone with an accurate memory or an encyclopedia at hand—the sort of errors authors and editors most dread, as opposed to the kinds of errors that are likely to go unrecognized or to be blamed on the printer. This problem of other less visible distortions is one that occurs to Freud, since he refers to it, in general terms, as follows:

> It has also occasionally happened that friends and patients, whose dreams I have reported, or have alluded to in the course of my dream-analyses, have drawn my attention to the fact that the details of the events experienced by us together have been inaccurately related by me. These again could be classified as historical errors. After being put right I have examined the various cases and here too I have convinced myself that my memory of the facts was incorrect only where I had purposely distorted or concealed something in the analysis. (1901, 220)

Thus Freud is never careless, though he does not identify or correct any of these historical errors in the dream book that his readers cannot recognize on their own. Better still, the purposive distortions or concealments are also blameless, because unconscious. The same may be wished by all writers: as a writer writing, one can only write Amen.

Nothing better illustrates Freud's dramatic combination of negative dream wishes and positive self-presentation than his treatment of ambition, since the various connivances of ambition are easily the most common secret meaning attributed to his own dreams. As I have indicated, citing Wittgenstein's remark, very few of the forty-odd personal dreams that Freud calls upon in his book have a sexual content, despite his growing belief that *all* dreams can be traced to some infantile sexual wish. The themes encountered in the specimen dream—professional responsibility betrayed by self-justification, competitiveness, and hostility toward colleagues—thread their way through *The Interpretation of Dreams* as a whole; and these themes are closely tied up with ambition, which emerges as a theme in its own right in the preamble to the second featured dream, that of his uncle with the yellow beard. "Once again this will involve me in a variety of indiscretions," he warns; "but a thorough elucidation of the problem will compensate for my personal sacrifice." The problem alluded to is the censorship, or what the chapter title refers to as "distortion." Freud again supplies a date for the dream, though this time an approximate date: sometime after he learned "in the spring of 1897" that two senior colleagues "had recommended me for appointment as *professor extraordinarius.*" Since a number of other qualified candidates of his acquaintance had been disappointed lately, however, Freud resigned himself to the uncertainty of his chances. "So far as I knew, I was not an ambitious man; I was following my profession with gratifying success even without the advantages afforded by a title. Moreover there was no question of my pronouncing the grapes sweet or sour: they hung too far over my head" (136–37).

This denial of ambition and refusal to be moved protect Freud's feelings and reputation against the secret meaning of his uncle with the yellow beard, the first of the dreams which reveal that he does harbor ambition and hopes for a professorship. In his account of this very brief and obscure dream (highly condensed, doubly displaced), Freud focuses on his reluctance to complete the interpretation: he does not want to entertain these wishes, does not expect to be preferred over anyone else. The interpretation reveals to him a wicked thought—that he regards his friends R. and N. as rivals for the professorship: "It made one of them, R., into a simpleton and the other, N., into a criminal, whereas *I* was neither the one nor the other." Both the preamble to the dream and Freud's sturdy resistance to interpreting it, however, reassure him and his readers (he is writing a book) that he understands well the correct attitude to be adopted toward ambition. As if this reassurance were not enough, he begins to ironize his interpretation and seems to depreciate the entire process.

> I was still uneasy over the light-heartedness with which I had degraded two of my respected colleagues in order to keep open my own path to a professorship. My dissatisfac-

tion with my conduct, however, had diminished since I had come to realize the worth that was to be attached to expressions in dreams. I was prepared to deny through thick and thin that I really considered that R. was a simpleton and that I really disbelieved N.'s account of the blackmailing affair. Nor did I believe that Irma was really made dangerously ill through being injected with Otto's preparation of propyl. In both these cases what my dreams had expressed was only *my wish that it might be so.* (140)

These remarks are preparatory to taking up one more feature of the dream, the feeling of warmth he felt for his uncle, but that feature is only the ingenious work of the censorship (the metaphor he is about to introduce) and does not alter the meaning.

Once the reader is acquainted with the theme of ambition, the secret meanings of Freud's dreams become less and less mysterious. "So once again I was wanting to be a Professor!" he jokes about the dream of Otto looking ill, with Professor R. in the wings (271). Biographers necessarily have to weigh the role of anti-Semitism in delaying the award to Freud, since anti-Semitism was a potent force in late nineteenth-century Vienna, and perhaps only for a Jew could the title of professor extraordinarius be in some sense a forbidden wish. Peter Gay (1988, 136–39) does a good job of sorting out the prevailing conditions that Freud faced, his personal reticence, and the reputation he sought for himself. Consideration should also be given, perhaps, to the publicizing of this wish in *The Interpretation of Dreams*, two years before his professorship was finally conferred by the state. That Freud can joke about it as well as keep finding it in his dreams says something of his resilience; that he commits these thoughts to print while awaiting a professorship may have been his indirect way of asking for it. There can be no question that the theme of ambition is two-sided. One does not need the dreams to wonder at the preamble "I was not an ambitious man," which in the original was cast in the present tense (Ich bin, soviel ich beweiss, nicht ehrgeizig). Before his twenty-eighth birthday Freud had written to Martha Bernays that "the world will not be able to forget my name just yet," and already he possessed a nice sense of paradox in this matter: "The trouble is I have so little ambition. I know I am someone, without having to be told so." A year later he wrote that he was destroying all his papers, except for her letters and those from his family: "As for the biographers, let them worry, we have no desire to make it too easy for them. Each one of them will be right in his opinion of 'The Development of the Hero,' and I am already looking forward to seeing them going astray" (1960, 105, 141). Whether the dream book figured in this program can be judged from these words to Fliess shortly after it was published, when its author was forty-three:

> For I am actually not at all a man of science, not an observer, not an experimenter, not a thinker. I am by temperament nothing but a conquistador—an adventurer, if you want it translated—with all the curiosity, daring, and tenacity characteristic of a man of this sort. Such people are customarily esteemed only if they have been successful, have really discovered something; otherwise they are dropped by the wayside. And that is not altogether unjust. At the present time, however, luck has left me; I no longer discover anything worthwhile. (1985, 398)

Whatever one says of Freud, he still seems unsure of himself after the masterpiece is done, though its rhetoric and universal propositions regarding dreams all bespeak a desire of fame.

Freud's denial of ambition is highly conventional, not meant to be deceptive. The same goes for the repeated, incessant assurances in the language of his interpretations that he takes an acceptable line on ambition. It is his modesty speaking, and modesty does not so much deny ambition as place it. Conventionally, female modesty is wrapped securely around sexuality; but as Ruth Bernard Yeazell (1991) has shown of the modern era, such modesty is best described as an approach to marriage, a prescribed narrative rather than a fixed virtue. Male modesty has just as conventionally been wrapped around ambition in the same era. Some glimpses of ambition or its suppression in Freud's childhood, afforded by the dream book and revered ever since in the history of psychoanalysis, may actually be seen as commonplaces of bourgeois upbringing. When, later in the book, Freud returns to the dream of his uncle with the yellow beard, he both typically exaggerates and modestly contains his ambition:

> If it was indeed true that my craving to be addressed with a different title was as strong as all that, it showed a pathological ambition which I did not recognize in myself and which I believed was alien to me. I could not tell how other people who believed they knew me would judge me in this respect. It might be that I was really ambitious; but, if so, my ambition had long ago been transferred to objects quite other than the title and rank of *professor extraordinarius*. (192)

That a peasant woman prophesized to Freud's mother that the infant would one day be a great man, and a cabaret poet declared in verse that the eleven-year-old would "probably grow up to be a Cabinet Minister," are pleasantries that few middle-class parents and children in this world are spared. Freud himself wryly comments that, since a handful of Austrian ministers had been Jews by that time, "every industrious Jewish schoolboy carried a Cabinet Minister's portfolio in his satchel" (193). Still more memorable in Freudian hagiography, because of the obvious materials for Oedipal construction, is the incident when he "disregarded the rules which modesty lays down and obeyed the calls of nature in my parents' bedroom while they were present." This misadventure at the age of seven or eight brought down the paternal reprimand "The boy will come to nothing" (Aus dem Buben wird nichts werden). Freud observes that "this must have been a frightful blow to my ambition, for references to this scene are still constantly recurring in my dreams and are always linked with enumeration of my achievements and successes, as though

I wanted to say: 'You see, I *have* come to something'" (216). But it would be surprising if some such blunder and reprimand did not afflict most children of that time and class, even if one or more generations in the United States have now been spared somewhat by the teachings of Benjamin Spock (1946), imbued with Freud's wisdom. Note how the euphemism "calls of nature" (Bedurfnisse) preserves the modesty of the adult writer and his readers. A book could be written on the decorousness of Freud's language even as, or especially when, he delves for that which might be seen as indecorous.

Male modesty is a measure of the distance between ambition and self-presentation. Freud is unmatched in his dramatization of this distance, hence second to no man in modesty. Approaching his principal dream of self-glorification—as opposed to mere ambition—he warns readers in advance that they must be prepared to be disgusted. This is the dream of the so-called open-air closet, which in British English is even more decorous than the original (ein Abort im Freien). The seat of this outdoor toilet is covered with shit, and Freud in his dream washes it clean. "*I micturated on the seat; a long stream of urine washed everything clean; the lumps of faeces came away easily and fell into the opening.*" Nor did he feel any disgust:

> Because, as the analysis showed, the most agreeable and satisfying thoughts contributed to bringing the dream about. What at once occurred to me in the analysis were the Augean stables which were cleansed by Hercules. This Hercules was I. . . . The seat (except, of course, for the hole) was an exact copy of a piece of furniture which had been given to me as a present by a grateful woman patient. It thus reminded me of how much my patients honoured me. Indeed, even the museum of human excrement could be given an interpretation to rejoice my heart. . . . The stream of urine which washed everything clean was an unmistakable sign of greatness. It was in that way that Gulliver extinguished the great fire in Lilliput—though incidentally this

brought him into disfavour with its tiny queen. But Gargantua, too, Rabelais' superman, revenged himself in the same way on the Parisians ... The fact that all the faeces disappeared so quickly under the stream recalled the motto: "*Afflavit et dissipati sunt,*" which I intended one day to put at the head of a chapter upon the therapy of hysteria. (468–69; see also 213–14)

Freud, it seems, has come a long way since pissing in his father's chamber pot as a child; at the same time he exuberantly portrays any self-glorification as both funny and disgusting. In a second phase of the interpretation, he punctures altogether this heroic and divine afflatus by recalling a more ordinary and modest ground for disgust with greatness. On the evening before the dream, though Freud felt he had given a listless lecture, a member of his audience insisted on joining him for coffee and proceeded to flatter him. "He told me, in short, that I was a very great man. My mood fitted ill with this paean of praise; I fought against my feeling of disgust, went home early to escape from him, and before going to sleep turned over the pages of Rabelais" (470).

Since Freud was not an ordinary man and rapidly achieved fame in middle life, his personal ambition—while always of interest—cannot provide to social historians a reliable sample of male middle-class attitudes. Still, his career before 1900, his abiding concern with ambition, and his determination to be modest are far from unsymptomatic of modern times. In a passage added to the book in 1911, he writes plainly of an experience shared by many in a society that requires individuals, particularly male individuals of the bourgeoisie, to maximize their personal efforts in whatever niche of the economy they occupy and, if possible, to find some new niche in which they can be yet more productive.

As a young doctor I worked for a long time at the Chemical Institute without ever becoming proficient in the skills which that science demands; and for that reason in my waking life I have never liked thinking of this barren and indeed humiliating episode in my apprenticeship. On the other hand I have a regularly recurring dream of working in the laboratory, of carrying out analyses and of having various experiences there. . . . Since those days I have become an "analyst," and I now carry out analyses which are very highly spoken of, though it is true that they are "*psycho-analyses.*" It was now clear to me: if I have grown proud of carrying out analyses of that kind in my daytime life and feel inclined to boast to myself how successful I have become, my dreams remind me during the night of those other, unsuccessful analyses of which I have no reason to feel proud. They are the punishment dreams of a *parvenu,* like the dreams of the journeyman tailor who had grown into a famous author. (475)

Freud may exaggerate a little his reputation at the time, yet polite self-advertisement and modest ambition here all but counter his thesis that a dream is the fulfillment of a wish: "the punishment dreams of a *parvenu*" would seem to contain within them a check to ambition. Freud's closing reference is to another self-made writer, Peter Rosegger. The dream of Rosegger's that he discusses and a comparison with a Grimms' fairy tale, "The Little Tailor, or Seven at a Blow"—for which Otto Rank is credited (477)—help to generalize the picture; and perhaps Freud is also recalling the humiliating apprenticeship of David Copperfield.

The discomfort of these laboratory dreams (which Freud treats collectively) causes him to hesitate as he reflects that they ought to represent wish fulfillment. But one of the dreams might be construed, he thinks, as a wish to be young again: "So I was once more young, and, more than everything, *she* was once more young—the woman who had shared all these difficult years with me" (476). For a dream about his son as a war casualty, added as a footnote in 1919, Freud finds a similar out: "Deeper analysis at last enabled me to discover what the concealed impulse was which might have found satisfaction in the dreaded accident to my son: it was the envy which is felt for the young by those who have grown old, but which they believe they have completely stifled" (560). The motive of such dreams, if Freud has perceived it correctly, is too commonplace to have needed concealment. Though *reality* may be said to forbid wishes to be young again, metaphorically "dreams" have opposed reality since ancient times. Every wish itself opposes reality in some sense. Often Freud is content to read back into his dreams some accepted wisdom, and quite naturally from the perspective of the middle class. The Count Thun dream and its analysis openly show this bias. One of Freud's most involved studies of personal ambition, the interpretation brings down a vigorous cold shower of modesty, as he writes that three episodes in the dream "were impertinent boastings, the issue of an absurd megalomania which had long been suppressed in my waking life and a few of whose ramifications have even made their way into the dream's manifest content (e.g. *'I felt I was being very cunning'*)" (215). Despite the intricacy of Count Thun, with its personal associations, Freud's allusions to Beaumarchais and to Mozart's Figaro lend the interpretation an air of appeal for class sympathy. When he returns to this dream later, he invokes Figaro again and contends—quite contrary to how he has explained the dream work before—that its absurdity was produced by a dream thought about something absurd, which he gives as follows: "It is absurd to be proud of one's ancestry; it is better to be an ancestor oneself" (434). Since Don Quixote came to the very same conclusion, this middle-class sentiment is at least as old as Cervantes (1605–15, 207).

Aside from professional life, one of Freud's earliest wishes was to visit Rome, a wish he failed to achieve until 1901, more than a year after the publication of *The Interpretation of Dreams.* The dreams about Rome in the book have also to do with ambition, and if anything these dreams have drawn more attention than the others, probably because they do not seem as trite as dreams of professional advancement. The Rome motif has been seen by Schorske, for example (1973), as an index of Freud's political position, and therefore of ambition in the grander sense of the just claims upon history of a Jew brought to Vienna with his family from Moravia in 1859–60. In discussing the fourth dream in the series, Freud writes of his boyish identification with the famous Carthaginian general Hannibal, as if in rebellion against the course of ancient history.

> Like so many boys of that age, I had sympathized in the Punic Wars not with the Romans but with the Carthaginians. And when in the higher classes I began to understand for the first time what it meant to belong to an alien race, and anti-semitic feelings among the other boys warned me that I must take up a definite position, the figure of the semitic general rose still higher in my esteem. To my youthful mind Hannibal and Rome symbolized the conflict between the tenacity of Jewry and the organization of the Catholic church. And the increasing importance of the effects of the anti-semitic movement upon our emotional life helped to fix the thoughts and feelings of those early days. Thus the wish to go to Rome had become in my dream-life a cloak and symbol for a number of other passionate wishes. (196–97)

As interpretation yields to reminiscence, there are no more affecting pages in the dream book. Of course, there is nothing singular about a boy identifying with a particular hero. Erikson (1968, 22–23), inspired no doubt by this very passage, reasons that the relation to a historical "type" is an important part of any child's sense of identity. But Freud continues, thoughtfully turning over his early memories until they subsume his father's memories of what it was like to be a Jew in still earlier times. He remembers a walk when his father told the story of another Saturday in the town where Freud was born. "'A Christian came up to me and with a single blow knocked off my cap into the mud and shouted: "Jew! get off the pavement!"' 'And what did you do?' I asked. 'I went into the roadway and picked up my cap,' was his quiet reply" (197). This was the interchange that made

Freud think of Hannibal's father, who made his son swear to take vengeance. In the loose associative arrangement of the dream book, the passage takes hold of the reader unforgettably before merging again with less memorable detail.

The road to Rome assumes a high literary path in Freud's *Non vixit* dream, the substance of which is professional life again. This is another laboratory dream, peopled by two colleagues from the laboratory days, one of them since deceased, plus Freud himself and his friend Fl., visiting town *"unobtrusively in July"* (421). Again there seem to have been political overtones, since Freud much later realized he had borrowed the Latin words from the Kaiser Josef Memorial in Vienna. In the dream itself, however, he was aware that he had the tense of the Latin verb wrong, because he was trying to say to Fl. that P. was no longer alive. As he uttered the words and stared at P., the revenant gradually faded away; and analysis shows that the dream has everything to do with killing looks. The sources include a look once directed at the dreamer himself when he was late to work and another metaphorically directed at another colleague by P. (484). In the interpretation of *Non vixit*, ambition has become just a little murderous. The main look that kills is Freud's own, but the usual disclaimer of ambition also suggests that he is virtuously protecting P. He cannot convey "the complete solution," Freud writes, because he is "incapable of doing . . . what I did in the dream": incapable, that is, "of sacrificing to my ambition people whom I greatly value" (422). The dream thoughts are at once hostile and affectionate, he decides. "In the scene in the dream there was a convergence of a hostile and an affectionate current of feeling towards my friend P., the former being on the surface and the latter concealed, but both of them being represented by the single phrase *Non vixit*" (423).

At this point, the interpretation creates its own literary associations, with a pleasing echo that demonstrates, among other things, how far Freud is from believing that science should view the evidence as something separate from the inquiry. The interpretation becomes the evidence, even as he sets it down on

paper, and not the semantics of the words but their cadence. Of
P. in the dream, Freud writes, "As he had deserved well of science I built him a memorial; but as he was guilty of an evil wish
(which was expressed at the end of the dream) I annihilated
him." Without a break he continues, directing attention to the
sentence he has just written:

> I noticed that this last sentence had a quite special cadence,
> and I must have had some model in my mind. Where was
> an antithesis of this sort to be found, a juxtaposition like
> this of two opposite reactions towards a single person, both
> of them claiming to be completely justified and yet not in-
> compatible? Only in one passage in literature—but a pas-
> sage which makes a profound impression on the reader: in
> Brutus's speech of self-justification in Shakespeare's *Julius
> Caesar*, "As Caesar loved me, I weep for him; as he was
> fortunate, I rejoice at it; as he was valiant, I honour him;
> but, as he was ambitious, I slew him." (423–24)

From this resemblance between his words of interpretation and
the language of Shakespeare's play, Freud decides that he was
Brutus in the dream, slaying P. because P. was ambitious, even
though he has just given "my ambition" as the motive. Nothing
before now has triggered anything Roman in the dream except
for the Latin expression *Non vixit* when *Non vivit* was meant,
but in searching for confirmation of his Shakespearean association, Freud hits upon the false information in the dream that his
friend Fl. (that is, Fliess) had been in Vienna in July, the month
of the calendar named for Julius Caesar.

"Strange to say," the next paragraph begins, "I really did once
play the part of Brutus" (424). When I first read this years ago,
I thought to myself—the way Freudians love to one-up Freud—
that he was wishfully misremembering Polonius, who informed
Hamlet and the court that *he* once played the part of Caesar ("I
was kill'd i' th' Capitol; Brutus kill'd me"): a theatrical joke quite
worthy of *The Interpretation of Dreams*. But it seems definitely
the case that Freud when he was fourteen recited Brutus's half of

the ballad sung by Karl Moor in Schiller's *Die Räuber*, a dialogue between the shades of Caesar and Brutus when they meet in the underworld; and William J. McGrath (1986, 83) has uncovered documentary evidence that he played Brutus again when he was fifteen, opposite Cassius in the quarrel scene from Shakespeare's play. The part of Caesar, in Schiller's ballad, was played by his nephew John—one year older than he—on a visit to Vienna from Manchester. As Freud (1985, 268) indicated to Fliess in 1897, he believed his relation to this nephew was formative of his later friendships, a theme which he enlarges upon in the interpretation of *Non vixit*. "We had loved each other and fought with each other; and this childhood relationship, as I have already hinted above, had a determining influence on all my subsequent relations with contemporaries" (424). The statement reaches well beyond dreams, obviously, and while this notion of the determining influence of childhood experience was not invented by Freud alone, it is clearly one of the important contributions of the dream book.

When he next refers to the relationship with his nephew, Freud calls it flatly "the source of all my friendships and all my hatreds" (472). Such an appeal to a formative experience has become commonplace today largely through Freud's influence: the Oedipus complex itself falls into this mode of explanation. He touches yet again on the nephew when he returns to the *Non vixit* dream later in the book:

> I have already shown how my warm friendships as well as my enmities with contemporaries went back to my relations in childhood with a nephew who was one year my senior; how he was my superior, how I early learned to defend myself against him, how we were inseparable friends, and how, according to the testimony of our elders, we sometimes fought with each other and—made complaints to them about each other. . . . My nephew himself re-appeared in my boyhood, and at that time we acted the parts of Caesar and Brutus together. (483)

The difficulty is in gauging Freud's need for this early model for friendship and enmity, as if such sentiments did not commonly arise between adults. Even his literary reversion to the model in school, in playing the part of Brutus, occurred nearly thirty years prior to the writing. It raises a Wittgenstein-like question of whether the relationship was a cause or an explanation more in the nature of a myth. The reference to "the testimony of our elders" suggests precisely the kind of family legends out of which early memories are usually composed. "My emotional life has always insisted that I should have an intimate friend and a hated enemy," Freud continues, and it seems that the experience insists on an explanation. "It has not infrequently happened that the ideal situation of childhood has been so completely reproduced that friend and enemy have come together in a single individual," he writes—"though not, of course, both at once or with constant oscillations, as may have been the case in my early childhood" (483). A childhood ideal (Kindheitsideal) apparently assuages a less comforting and more reproachable adult reality. Students of Freud's life will likely conclude that he means his relationship with Breuer or Fliess—and writes proleptically of tense personal relationships still ahead. In nearly his last word on *Non vixit*, Freud notes that the dream expressed a sense of good fortune, that in losing friends he had been able to find others. But his interpretation continues to deposit any feelings of hostility back in childhood, roughly forty years ago, when he could hardly be to blame for them.

> My satisfaction at having found a substitute for these lost friends could be allowed to enter the dream without interference; but there slipped in, along with it, the hostile satisfaction derived from the infantile source. It is no doubt true that infantile affection served to reinforce my contemporary and justified affection. But infantile hatred, too, succeeded in getting itself represented. (486)

Thus infantile affection reinforces present affection, but hatred is isolated in the dream and confined to the past: there is no present hatred for infantile hatred to reinforce. Any hatred was childish and inspired by persons without any active relation to the dreamer, who harbors no improper feelings for anyone at present. An interpretation with this capacity for attributing hostility to the repressed or scarcely remembered wishes of childhood is obviously itself wish fulfilling. It is consoling with respect to the present and therefore a suitable move in psychoanalytic therapy. Yet one wonders that Freud would make such a deep thing of these socially unacceptable feelings, when he meant no harm in any case.

The literary associations of *Non vixit* are probably as important as the personal relations. Schiller's tragedies were prized by German culture much as Shakespeare's were in England, and in truth Schiller mediated a good deal of Shakespeare to German speaking people. *Die Räuber* may have made as lasting an impression on a Viennese schoolboy as the equally youthful—that is, filial—tragedy of *Hamlet.* As a child Freud would have known that the heroine and sole woman of the cast, Amalia, had the same name as his mother (1901, 24). The configuration of the plot, with the aging father, rival sons, and Amalia, was already suggestive of the primal myth of the family Freud would elaborate in *Totem and Taboo* (1913, 140–61). Schiller's first play also drew on the story of the prodigal son and touched on the story of Jacob and Joseph in the Bible, besides being filled with the heady Rousseauism of its time. McGrath, who has treated at length the influence of the play on Freud, finds its political inspiration behind "the concept of psychoanalytic freedom." The very coincidence of Freud's own family concerns and the play's starker confrontations, which breach not only the family but the law, may have helped radicalize the founder of psychoanalysis, McGrath thinks: "The drama served to translate emotionally charged familial relationships into passionately held political views" (1986, 294, 70). I do not doubt the influence of *Die Räuber* on Freud's myth making, but his politics were more circumscribed than this conclusion implies. One would have to ask how the passion in question shows itself. "In politics," Peter Gay judges, "Freud was a man of the center" (1988, 387; also 547–50). True, Gay is addressing especially Freud's politics after the First World War; but already in *The Interpretation of Dreams* the writer's self-presentation is one of detachment: detached from politics, detached from passion. In fact, for all the marvelous intimacy, the memories lingering in the air of discovery, the dream book is also very much detached from youth.

The significance of Schiller is, precisely, in the filial relation. As Freud interprets his *Non vixit* dream, the cadence of his own words reminds him of the oration of Shakespeare's Brutus. This association he then caps with a different sort of memory—"I really did once play the part of Brutus"—which proves to be the script written by Schiller for Karl Moor in *Die Räuber*. In Shakespeare's *Julius Caesar* there are not even two generations carefully delineated: Brutus and Cassius are almost of an age with Caesar, Antony and Octavius somewhat younger. The question for the conspirators is whether they will allow anyone to be first among equals, let alone be crowned. Shakespeare knew, from Plutarch, the legend that Brutus may have been a son of Caesar; but he chose not to feature it in his play, which points more daggers at Caesar's ambition than anyone else's, while it preserves a political neutrality. Schiller's tragedy, on the contrary, is first and foremost about two generations of the same family. The misunderstandings of father and son are usefully divided between the rival brothers, the outlaw hero Karl and the dissembling villain Franz. In fact, Schiller imports wholesale from *King Lear* the rivalry of Edgar and Edmund, including Edmund's apostrophe to nature—a familial plot quite foreign to *Julius Caesar*. And by accepting the legend that Brutus was the son of Caesar, the playwright can justify the ballad written for act 4, performed by Karl Moor in the play and divided for school presentation between Freud and the nephew one year older than he. The ballad (a translation is reprinted by Grinstein, 1968, 300–301) has five stanzas, three for the shade of Brutus and two for the shade of Caesar: thus Brutus has the last word. Caesar reproaches Brutus for killing his own father, and Brutus rejoins that they are equals.

In *Non vixit*, according to Freud, professional rivalries in the laboratory get out of hand: his feelings toward his friend P. were both affectionate and hostile, so he proceeded to annihilate P. In the extended discussion of the dream, he remarks that P.'s desire to have his superior "out of the way might have an uglier meaning than the mere hope for the man's promotion. Not unnaturally, a few years earlier, I myself had nourished a still livelier wish to fill a vacancy." Apparently he is not writing solely of his own experience, since he immediately generalizes in these curious terms: "Whenever there is rank and promotion the way lies open for wishes that call for suppression" (der Weg für der Unterdrückung bedürftige Wünsche eröffnet). The force of wishes *calling for* suppression is puzzling, open to any number of interpretations. The constraint in question might be affection, politeness, morality, prudence, or the sanctions of the criminal code. The parameters, I would stress, are bound to be social, if only because he is writing a book rather than a private document. Next, without a break, comes this sentence: "Shakespeare's Prince Hal could not, even at his father's sick-bed, resist the temptation of trying on the crown" (484). From "wherever there is rank and promotion" the scene reverts to the family, to royalty, to Shakespeare, and very likely to the thought of Freud's own father's deathbed. The one thing he leaves out is the memorable opening line of Henry IV's long rebuke of Prince Hal on that occasion: "Thy wish was father, Harry, to that thought"! If Freud could recall the cadence of Brutus's self-justification, it is surprising he did not remember this witty put-down of wish fulfillment. Of course Brutus was the "son," and King Henry the father. Freud really did play the part of Brutus, even as the nineteenth century, it might be urged, took the part of sons in its fairy tales, its factories and offices. A wish was father to this state of affairs and so also to Freud's theory.

The theme of modest ambition—*the* autobiographical theme of the dream book—ought to be regarded very closely in the section entitled "Dreams of the Death of Persons of Whom the Dreamer Is Fond" (Traume vom Tod teurer Personen). That section, inspired by Freud's feelings about his father and inspiring the Oedipus complex in turn, exhibits the same mixture of affection and hostility and the same potential for murder as the more numerous accounts of professional rivalries. Patently the Oedipus complex is a theory conceived from a wish, or more exactly a wish in two motions: to put the father out of the way and to possess the mother sexually (this is of course a male child's story). If there is one great myth that people associate with Freud, whether they have read his writings or not, it is this; and for some people it becomes a scientific explanation of what they feel. But so closely is the Oedipus complex tied up with childhood, with family, and with literature, that we do not immediately think of it, in either or both of its aims, as a special category of ambition. If anything, in psychoanalytic theory, ambition is supposed to originate with the earliest wishes characterized by the Oedipus complex. In my judgment, the horse is ambition and the cart is the Oedipus complex, and the evidence of the dream book suggests that Freud put the cart before the horse for the same reasons of comfort and correctness that he discovered in his childhood relation to his nephew "the source" of all his friendships and all his hatreds.

In the Western literary heritage tragedy is seldom separable from ambition, and the two great literary models that Freud cites in this section of *The Interpretation of Dreams* are no exception, for both Sophocles' *Oedipus Rex* and Shakespeare's *Hamlet* represent dynastic struggles and failures of leadership. But much more significantly, in these renowned pages of the dream book Freud discourses very plainly of the relations of parents to children and most notably of sons' impatience to replace their fathers in the present day. Nothing is at all unconscious about such facts or their history:

Let us consider first the relation between father and son. The sanctity which we attribute to the rules laid down in the Decalogue has, I think, blunted our powers of perceiving the real facts. We seem scarcely to venture to observe that the majority of mankind disobey the Fifth Commandment. Alike in the lowest and in the highest strata of human society filial piety is wont to give way to other interests. (256)

Here Freud instances the behavior of Kronos and of Zeus, as providing "an unpleasing picture" of the kind of thing he means and generalizes that when paternal power was unconstrained by law, the son must "have found himself in the position of an enemy" impatient to replace him. These are the sentences of the passage most often quoted—as they were by the author himself because of his error about Zeus. But Freud returns once more from ancient times to the present, with the following words:

Even in our middle-class families fathers are as a rule inclined to refuse their sons independence and the means necessary to secure it and thus to foster the growth of the germ of hostility which is inherent in their relation. A physician will often be in a position to notice how a son's grief at the loss of his father cannot suppress his satisfaction at having at length won his freedom. In our society to-day fathers are apt to cling desperately to what is left of a now sadly antiquated *potestas patris familias*. (257)

In writing thus calmly of the social parameters of the Oedipus complex, Freud does as much as Gilles Deleuze and Félix Guattari (1972) have since done—with so much more ado—to qualify his theory.

Freud does not neglect the daughter of the family entirely. His emphasis on sexuality is such that he could hardly leave her out, in this early formulation of the Oedipus complex around the subject of dreams. He observes, shrewdly enough, "that dreams of the death of parents apply with preponderant frequency to the parent who is of the same sex as the dreamer. . . . It is as though—to put it bluntly—a sexual preference were making itself felt at an early age: as though boys regarded their fathers and girls their mothers as their rivals in love, whose elimination could not fail to be to their advantage" (256). No doubt there is some truth here, because—other considerations apart—attraction is usually to the opposite sex. As grounds for a theory, however, it makes any murderous wishes subordinate to the sexual, which of course need not be the case in dreams of professional advancement. As so often with Freud's inferences, he neglects equally obvious, or more obvious, explanations and settles for the one he likes—in this case, unconscious incestuous thoughts. The choice is a little scandalous, as the aside "to put it bluntly" (grob ausgesprochen) indicates and as reference to these thoughts as "monstrous" (ungeheuerlich) in the following paragraph insists. Yet Freud can think of other motivations for same-sex rivalry, as his sermon on the Fifth Commandment in the same section reveals. In fact, after he takes up the social relations of fathers and sons, he duly provides this separate sentence on mothers and daughters: "Occasions for conflict between a daughter and her mother arise when the daughter begins to grow up and long for sexual liberty, but finds herself under her mother's tutelage; while the mother, on the other hand, is warned by her daughter's growth that the time has come when she herself must abandon her claims to sexual satisfaction" (257). Though the remark studiously confines a woman's interest to sex and her sexual opportunities to youth, any such commonplace identifications of mother and daughter are intuitively more likely than infantile wishes for a sexual relation with the full-grown mate of the same-sex parent. There is much to be learned from such casually dropped wisdom in Freud's book. Since he writes touchingly and repeatedly of adult male wishes for promotion, is it not possible that childish wishes are preponderately of the same order? That one of the most persistent and fertile wishes of children is the wish to grow up? There can still be room for scandal here, as in the laboratory. As D. W. Winnicott, who has certainly learned from Freud, writes, "Growing up means taking the parent's place. *It really does.* In the unconscious fantasy, growing up is inherently an aggressive act" (1971, 144). And as Freud's examples hint, the aggression need hardly be unconscious.

Grown-up manners forbid aggression, however, and teach one not to care more deeply for a professorship than kings like Henry IV care for power and riches. Freud had lived four decades when he composed his masterpiece: suppose one takes his adult experience—even those fragments delineated in the book—as a way of accounting for the theory, and suppose one views modest ambition as a fixed persuasion and way of thinking *prior* to the Oedipus complex. How did Freud know that his interests and P.'s or Fl.'s interests were opposed? Or that fathers sometimes stand in the way of their sons' independence? Freud obviously experienced ambition before he made his discoveries, and just as ambition and affection cross in the laboratory dreams, so they cross in dreams of the death of parents. In each case, modesty—both good manners and the way one regards oneself—prefers to deflect hostility onto childhood. Given the formality of bourgeois mores, it may even be advantageous for men to think affection childish as well. The Oedipus complex, as Freud came to call it (1910, 171), is precisely the myth that assigns disloyal or unsocial feelings to childhood, even though the obvious grounds for entertaining murderous wishes are experienced as an adult. When one reads skeptically Freud's pages on dreams that kill, his careful back-dating of such wishes is apparent:

> If anyone dreams, with every sign of pain, that his father or mother or brother or sister has died, I should never use the dream as evidence that he wishes for that person's death *at the present time.* The theory of dreams does not require as much as that; it is satisfied with the inference that this death has been wished for at some time or other during the dreamer's childhood. I fear, however, that this reservation will not appease the objectors; they will deny the possibility

of their *ever* having had such a thought with just as much energy as they insist that they harbour no such wishes now. (249–50)

The idiom reveals that theories, too, fulfill wishes, since this one is satisfied (sie begnügt sich) with what it can get. Freud is being a little ironic, perhaps; but if so, the irony seizes on just that degree of adult aggression that the theory is said not to require. What is the point of chiding his readers, if he does not believe that they do harbor such wishes now? This last sentence portrays Freud as the solitary proponent of what is only reasonable to suppose; but what is reasonable, once again, are some feelings of aggression among members of the family. The Oedipus complex soon emerges from *The Interpretation of Dreams* to become the centerpiece of psychoanalytic explanation, a powerful myth; and since, as Freud introduces it, psychoanalysis resembles a science, the myth passes as its etiology. But if one views the unconscious wishes of childhood from this social perspective, also available in the dream book, the Oedipus complex, instead of an etiology, ought to be regarded as an apologetics.

I have sought to make this point before, with the help of Dickens (1987, 156–72). Often something similar is done by Freudians who, empowered by the dream book itself, never hesitate to pry out the motives of the master. From this officially sanctioned approach to case study, it has become perfectly proper—almost a professional challenge—to supply the wishes from which a new narrative about Freud can be constructed. Here are a few of Marianne Krüll's speculations, for example: "The replacement of the seduction with the Oedipus theory thus enabled Freud to examine his own childhood without having to blame his parents for his neurosis. According to the new theory, that neurosis was caused by his own forbidden desires. Nor did he have to blame himself for these desires, for they were universal" (1979, 68). The bent of Krüll's entire book, which contains much useful information, is to trace psychoanalysis to Freud's own relationship with his father. But I do not find Jacob Freud so singularly a failure or so weak as Krüll makes him out to be—though he was forty years older than Freud, to be sure, and eighty when he died in 1896. Freud tells the story of Jacob's hat being knocked off by a Christian not to call attention to his father's weakness but to express his own boyish hope to become a hero, for a second or third generation of emancipated Jews. It was Jacob, remember, who first told him the story. There can be little question that Sigismund, later Sigmund, became that hero in middle life and was a far more able, more productive man than his father. Yet Freud was not the only remarkable son of a not so remarkable father, and his filial feelings were not entirely unprecedented. Pleasingly, Ernest Jones reports that Freud described his father "in rather Micawber-like terms as being 'always hopefully expecting something to turn up'" (1953, 1.2). At least to Jones, *David Copperfield* suggested a model for accommodating affection and hostility, and vastly different destinies of father and son, in the figure of Micawber.

Freud came to think of the Oedipus complex as older than recorded history. If he was right, then the question to ask is why human genius waited until the end of the nineteenth century to celebrate its force. My guess is that this new awareness followed from the commercial and industrial revolution that demanded of each male participant, at least, a maximization of lifelong effort. Not Caesar but every citizen was to be ambitious, and necessarily so if mass production and a market economy could be made to work. But small boys still grew up in families, still took many years to match the physical stature of their fathers, and meanwhile generally honored their fathers and mothers as they were expected to. In short, a contradiction arose between the demands of the economic system and the demands of the bourgeois family. At the end of his life, Freud very movingly expressed his sense of rivalry with his father, a feeling not derived from infancy but from their diverging careers: "It seems as though the essence of success was to have got further than one's father, and as though to excel one's father was still something forbidden" (1936, 247).

Similarly, the modern individualistic political ideal deserted fathers as it did true patriarchy. Freud thought and wrote of society very much within the contractual theory of both Hobbes and Rousseau. Repression may be as old as civilization, but an intellectual understanding of repression is, practically speaking, no older than contract theory. Traditional authority need not inspire guilt feelings at all: disobey a king or an old-time patriarch and you *are* guilty. But under a government that is in theory contractual, repression becomes in principle self-repression. Although fathers may be outlived or on occasion may even forgive their sons, the social contract cannot be outlived and is endlessly unforgiving, because its authority over its subjects is self-imposed. Political theorists may know this, but how it feels to live the contract was first importantly described by novelists, I believe, beginning notably with Scott in the early nineteenth century (1992a, 88–99; 1992b). Freud recognized that repression had a certain history, for in commenting in the dream book on "the changed treatment of the same material" from *Oedipus Rex* to *Hamlet*, he remarked on "the secular advance of repression in the emotional life of mankind" (264). He also seems to have reflected while composing his masterpiece that the modern political and economic insignificance of fatherhood may have worsened tensions between parents and children by the Victorian age. "In our society to-day fathers are apt to cling desperately to what is left of a now sadly antiquated *potestas patris familias*" (257). Contrariwise and accordingly, the institutionalizing of ambition promoted professional and filial modesty in the modern era.

In the dream book Freud at once plays down and reveals his ambition. Here the significance of wishfulness is apparent, in this parallel with a difference: whereas success is the fulfillment of ambition, a dream is the fulfillment of a wish. Freud modestly adopts the right attitude toward ambition yet entertains the most scandalous aggressions, in dreams that need not stop at murder. He thereby entertains in every sense of the word: wishfulness is that passive mode of enjoyment to which active intent is constantly referred. And that which he accomplishes for himself, he accomplishes for his readers as well: troped as wishfulness, ambition need not be denied or disguised by anyone. Best of all, because this is a book of dreams and not deeds, it can perform the deed in question—attain that success, warranted by ambition, which is due to Freud's masterpiece.

CHAPTER FOUR

"It Had Been Possible to Hoodwink the Censorship"

I N THE DREAM BOOK, the insistence on secrets supports the censorship: dreams have a secret meaning; the censorship is in the business of keeping secrets. The censorship figures in the formal conclusion of the book, with its affirmation of "the two psychical systems, the censorship upon the passage from one of them to the other" (607). Freudian censorship is very much the invention of *The Interpretation of Dreams.* Freud had employed the word (Zensur) casually in *Studies on Hysteria* (1895, 269, 282) and an early paper, "The Neuro-Psychoses of Defence" (1896, 182, 185), but in the dream book he expressly introduced it as a metaphor. Well before the end, it became an institution of the mind that he could scarcely do without, and it remained so in the later writings. The censorship has since become part of the popular lore of psychoanalysis, for which the connotations of the original metaphor have done good service. For all that, the concept is difficult to pin down.

Freud's dream of his uncle with the yellow beard in chapter 4 sets the scene: "My dream thoughts had contained a slander against R.; and, in order that I might not notice this, what appeared in the dream was the opposite, a feeling of affection for him" (141). This is the observation that triggers the idea of the censorship, and significantly the words "in order that I might not notice" (damit ich diese nicht merke) already imply a purposive idea such as motivates dream wishes themselves: a purpose of keeping as well as revealing secrets is already unconsciously at work. Freud then proceeds:

> I will try to seek a social parallel to this internal event in the mind. Where can we find a similar distortion of a psychical act in social life? Only where two persons are concerned, one of whom possesses a certain degree of power which the second is obliged to take into account. In such a case the second person will distort his psychical acts or, as we might put it, will dissimulate. The politeness which I practise every day is to a large extent dissimulation of this kind; and when I interpret my dreams for my readers I am obliged to adopt similar distortions. The poet complains of the need for these distortions in the words:
>
> > Das Beste, was du wissen kannst,
> > Darfst du den Buben doch nicht sagen.
> >
> > (141–42)

For understanding the dream book and its appeal, no passage is more important than this deliberate comparison to social life. The dramatic potential of dream interpretation is confirmed directly, since only the interaction of two unequal persons affords a comparison to the internal event. Because the power of the second person in Freud's figure is apparently customary and communal, it would seem that dream wishes are the destined losers and the defense against them eminently social (no other power than social is suggested). The three examples indicate roughly the kinds of distortion Freud has in mind, while each bristles with a veiled allusion to the writing he has in hand: everyday politeness (insofar as polite people say one thing and believe another); his own devices for protecting himself from his readers (here cast as more powerful than the writer); and, by means of the quotation from Goethe's *Faust*, part 1, lines 1840–41, things that grown people do not tell children (here readers are reduced to *Buben* and the writer is grown up). In these interactions, the expressive agent and likely loser keeps control over the expression, with a view to the reaction of the other party.

Strictly speaking, the censorship does not appear by name until the following paragraph, as a more specialized analogy is brought in to support this of the social life:

> A similar difficulty confronts the political writer who has disagreeable truths to tell those in authority. If he presents them undisguised, the authorities will suppress his words. . . . A writer must beware of the censorship, and on its account he must soften and distort the expression of his opinion. . . . The stricter the censorship, the more far-reaching will be the disguise and the more ingenious too may be the means employed for putting the reader on the scent of the true meaning. (142)

In this version, the first party still possesses the initiative, but through a kind of bureaucratization the second party apparently sets the rules or singles out that which may not be written; therefore whatever sense of control there is has been externalized and dramatized, or has become distinctly a different interest. The new analogy has also politicized the interaction, so that the suggestion of individual versus social interest tends to become radical minority versus entrenched state power. With his characteristic eagerness, Freud no sooner pens the analogy than he mines it for whatever etiology it may yield. "The fact that the phenomena of censorship and of dream-distortion correspond down to their smallest details justifies us in presuming that they are similarly determined" (143). An analogy justifies nothing of the kind, but for Freud the two parties to the social relation correspond to "two psychical forces" or "currents or systems," one of which produces the wish behind the dream and the other of which—"exercises a censorship." Since he adopts the metaphor in the very process of examining its usefulness, it seems doubtful that he is doing anything more than continuing to write metaphorically: that is, describing the operation of the mind as an interaction between a certain self-expressiveness and the realities of social communication. Yet he does not hesitate to invest the second force, current, or system with far-reaching authority.

> It remains to enquire as to the nature of the power enjoyed by this second agency which enables it to exercise its censorship. When we bear in mind that the latent dream-thoughts are not conscious before an analysis has been carried out, whereas the manifest content of the dream is consciously remembered, it seems plausible to suppose that the privilege enjoyed by the second agency is that of permitting thoughts to enter consciousness. (144)

One reason the metaphor has been chosen, probably, is the authority of the state that it brings to bear. Constraints of mere politeness and custom do not have such regulated being. At the same time, a subtext continues to portray the writing of the dream book itself as a struggle against what the readership will tolerate.

Censorship held the day, and it still holds the day in high psychoanalytic theory and common parlance. No idea introduced in Freud's masterpiece would appear to be more decisive, and without question the institution has lent a political coloring to our picture of mental life. Still, as far as the theory goes, this political suggestion can be narrowly misleading. Observing that Freud's motto for the book, from Virgil, had been previously used for a socialist work by Ferdinand Lassalle, which Freud had read, Carl Schorske (1973) demonstrated that *The Interpretation of Dreams* has more than one tie with youthful radicalism. But a book's origins do not necessarily typify the story it tells. In practice, when Freud writes of the dream censorship, he is forced to rely on his initial, more general metaphor of the distortions common to social life. Many instances have to do with the censoring of sexual thoughts not usually communicated to other people; and in the case of his own dreams, the censorship is usually concerned with some unworthy feeling instilled by ambition. Even when a dream has a definite political content, as in

Count Thun, the censored material is socially rather than politically unacceptable.

A passage in the "Affects in Dreams" section replicates the terms in which he first broached the subject, and in such a way that conscious or unconscious control scarcely comes into it:

> In social life, which has provided us with our familiar analogy with the dream-censorship, we also make use of the suppression and reversal of affect, principally for purposes of dissimulation. If I am talking to someone whom I am obliged to treat with consideration while wishing to say something hostile to him, it is almost more important that I should conceal any expression of my *affect* from him than that I should mitigate the verbal form of my thoughts. . . . Accordingly, the censorship bids me above all suppress my affects; and, if I am a master of dissimulation, I shall assume the *opposite* affect—smile when I am angry and seem affectionate when I wish to destroy. (471)

It may be noted that here Freud writes in full awareness of feelings that he elsewhere contends would appear only in disguised form as he sleeps. When something that in theory ought to be concealed appears, as in his second discussion of *Non vixit*, he may ask outright, "But what had become of the dream-censorship? Why had it not raised the most energetic objections against this blatantly egoistic train of thought?" (485). I shall not pause for his explanation but merely note what sort of thing should have been censored: immodest thoughts (Gedankengang der rohesten Selbstsucht) rather than sedition, say. Thus every now and then Freud expresses puzzlement about a dream in plain text rather than in code, but the problem may have more to do with his rash insistence on disguise than with the nature of the censorship. What the latter cannot do without is a knowledge of the kinds of thoughts that are acceptable and the kinds that are not. As I shall argue, Freud has no grounds other than social life for knowing what the censorship within him—or within anyone else—might censor.

In the dream book, censorship works so closely with displacement that confusion is possible on this score also. The two important functions of the dream work are clearly condensation (Verdichtung) and displacement (Verschiebung). "Dream-displacement and dream-condensation are the two governing factors to whose activity we may in essence ascribe the form assumed by dreams" (308). Condensation allows any number of latent thoughts to appear in the dream under a single representation and is an aspect of the dream work that, if it did not exist, would surely be created by the technique of free association in the process of interpretation. Displacement allows any latent thought to appear in the dream under a different guise and thus accounts for endless differences between the latent and manifest contents. Yet Freud often refers to the latter as simply the alteration in the dream brought about by the censorship. In the section on secondary revision, for example, instead of "displacement" after condensation he uses in quick succession the phrases "necessity for evading the censorship" (Zwang der Zensur auszuweichen) and "censorship imposed by resistance" (Widerstandszensur) (499). It seems that displacement, or the dream work generally, provides only an operating definition of the gap between latent and manifest contents and therein cannot satisfy Freud's usual mode of explanation: he seeks some kind of purposive function, which the censorship provides. Condensation and displacement as "the two governing factors" (die beiden Werkmeister) might be better translated as the foremen in charge of manufacturing dreams. Then the censorship, by virtue of its bureaucratic overtones, may be needed to direct these foremen.

Displacement often does seem to be in the service of the censorship. In his too-brief section on displacement, Freud notes that dream distortion is not a new subject in the book: "We traced it back to the censorship which is exercised by one psychical agency in the mind over another. Dream-displacement is one of the chief methods by which that distortion is achieved. *Is fecit cui profuit.*" This legal adage ("He who profited did it") insists that the censorship is as motivated as any dream wish, though Freud's irony leaves uncertain *how* it profits. Displacement, he writes, "comes about through the influence of the same censorship" (308). Similarly, any displacement can "serve the ends of the censorship" (471), or something important in the dream thoughts is reduced to indifference in a dream "for reasons of censorship" (589). This notion of displacement as the means but censorship as its authority still obtains in the *New Introductory Lectures* when Freud returns to the subject for the last time: "Displacement is the principal means used in the *dream-distortion* to which the dream-thoughts must submit under the influence of the censorship" (1933, 21). It is still difficult to grasp at times whether dreams oppose the censorship or cooperate with it, as Freud's own idiom veers back and forth between these alternatives. Sheer theory building forces more and more responsibility on the censorship: "There can be no doubt that the censoring agency, whose influence we have so far only recognized in limitations and omissions in the dream-content, is also responsible for interpolations and additions in it" (489). As the said agency (zensurierende Instanz) takes on the added suggestion of a court of justice, the metaphor becomes still more accommodating.

Yet the censorship as conceived is far from autonomous. Like a true bureaucracy, although it may hand out directives it also follows orders. The censorship that purposively guides displacement is often driven by resistance or repression, Freud seems to believe. At the end of the section on displacement, he writes that any "elements of the dream-thoughts" that appear in a dream *"must escape the censorship imposed by resistance"* (Z e n s u r d e s   W i d e r s t a n d e s, 308). Thereafter in the book Freud tends to combine these words (Widerstandszensur), while Strachey understandably continues to use the English phrase, "censorship imposed by resistance" (499, 542, 563). If "displacement" were not already available as an operative term, censorship might be thought of as a specific motion of resistance, and perhaps it still should be: again, it seems to provide a purposive edge that displacement lacks. (Less patient readers may feel that Freud defines these terms by means of one another and confuses them freely because they have no empirical base.) Another possibility is that the censorship can be regarded as a tentative form of repression. The general fact of repression is obviously fundamental for psychoanalysis. But if Freud is to interpret dreams or construct any plausible narrative from his patients' free associations, it becomes highly useful to detect tentative and particularized acts of repression that can be examined and disposed of. Repression will always be with us, but the censorship can be circumvented, the way a writer conceals "his objectionable pronouncement beneath some apparently innocent disguise" (142). Repression can always be assumed, whereas the censorship supplies fresh materials for a hermeneutics of suspicion. Freud gives some support to this tentative quality in his paper "The Unconscious," where he at one point refers to censorship as "a kind of testing" of ideas before they can pass from the unconscious (*Ucs.*) to the preconscious (*Pcs.*) and to consciousness (*Cs.*) (1915, 173). But by then, the metaphor institutionalized by the dream book suffers from severe abstraction:

> In discussing the subject of repression we were obliged to place the censorship which is decisive for becoming conscious between the systems *Ucs.* and *Pcs.* Now it becomes probable that there is a censorship between the *Pcs.* and the *Cs.* Nevertheless we shall do well not to regard this complication as a difficulty, but to assume that to every transition from one system to that immediately above it (that is, every advance to a higher stage of psychical organization) there corresponds a new censorship. (1915, 191–92)

Little rationale for the censorship remains at this level of abstraction except the necessity for these frontiers of concealment in the mind. But that is another way of saying that Freud's theory remains fascinated with secrets, which it can promise or threaten to reveal.

The metaphor seized upon in chapter 4 of *The Interpretation of Dreams* produced (or reproduced) more difficulties than Freud deserved. Elsewhere I have suggested—not altogether mischievously—that he might better have drawn a comparison to blackmail (1985, 337–77). He possibly chose the metaphor to balance his predilection for a highly energized unconscious. In chapter 7, when the psychology from his abandoned project reestablishes itself in the dream book, one senses that even the theory of wish fulfillment may have originated from this conviction about the nature of unconscious ideas. When Freud asserts "that dreams must be wish-fulfilments, since nothing but a wish can set our mental apparatus at work" (567), for example, it would seem that "wish" is nothing but a technical equivalent to something like force in physics and that the previous long chapters, with their wealth of theorizing and illustrative materials based on more everyday meanings of "wish," have gone for nothing. The censorship has been called upon to meet force with force, perhaps, and thus confirms the existence of two psychical systems in a highly dramatic and humanly understandable way. The underlying physics in chapter 7 may have contributed something all along to the element of coercion inherent in "censorship." Paul Ricoeur, who thinks the word "well chosen," notes that "censorship alters a text only when it represses a force, and it represses a forbidden force only by disturbing the expression of that force" (1970, 92–93). Still, the very success of the censorship idea traps Freud in two counterintuitive theses about memory: that no memories are ever lost completely, since they are present in the unconscious; and that unconscious memories continually strive to become conscious—the "upward drive" of the unconscious as Freud called it at the close of his life (1940, 179). Both assumptions are very strenuous: they render analysis a heroic endeavor, for the materials are unlimited and there must be constant mental activity of which we are unaware. If unconscious processes are "indestructible" and "in the unconscious nothing can be brought to an end, nothing is past or forgotten" (577), obviously there can be no such thing as simply forgetting. Though it might seem that wishes we have repressed would be the last to surface on their own accord, "these unconscious wishes are always on the alert, ready at any time to find their way to expression" (553). Very likely Freud was wrong on both these scores. J. Allan Hobson—who is not altogether unsympathetic, given that few neurobiologists are interested in dream interpretation at all—claims that our earliest memories and others become "irretrievably lost," and though Hobson does not cite what studies he refers to in support of this conclusion, his own theory of dreaming emphatically "makes no distinction between manifest and latent content" (1988, 44, 68). A readiness to get along without the censorship is also apparent in Fisher and Greenberg's critique of Freud (1985, 30–46).

To do away with the censorship, however, would be to do away with the dream book in its most enjoyable part. As Ricoeur proposes, a slogan for psychoanalysis might be "*Guile will be met with double guile*" (1970, 34). Without the guile of secret wishes and the censorship there would be no occasion for the double guile of analysis, and no reward for readers from the process of decoding what has putatively been coded. Still, the concept is mired in such difficulties in the dream book that it scarcely can be rationalized. The censorship does what it does, so that the analysis can be performed. Just as no adequate idea of the range of wishes that interest Freud can be attained except by reviewing the series of interpretations his masterpiece provides, the censorship cannot adequately be defined except case by case in the same series. To prepare such a survey, needless to say, would require a book both tedious and long, but let me make two predictions of what the survey would reveal (before going on in a moment to examine closely two samples). First, Freud has no other means of knowing what the censorship, in a given instance, has censored than a general knowledge of what is socially permissible to communicate in the culture that surrounds him.

"Nor will anyone regard it as a chance coincidence," he states at the end of chapter 4, "that the interpretation of these dreams has brought us up each time against topics which people are loth to speak or to think" (159). Well, this is not a coincidence: the topics unconsciously censored *are* the topics people dislike thinking or speaking about—though topics that, almost by definition, the same people enjoy as scandal. Second, the censorship typically blunders: while guilefully occupied with some wish, it has not calculated on the double guile of interpretation. In every interpretation that Freud presents, the censorship has failed, because once a displacement has been identified, so has the hidden wish. Thus the dream wish that was initially cast as a loser always wins, insofar as it becomes known to the author and readers of the dream book, and possibly to others—to patients or colleagues, for example—before the book is written. Double wish fulfillment thus becomes the rule: not only is each dream the fulfillment of a wish but each interpretation fulfills a wish by unveiling the same. The repeated falls of the censorship I read as the comical romance of the dream book, in its capacity as frame story.

So let us look at two cases in the kind of detail that Freud's storytelling warrants. Both are short enough to present in their entirety. One is about a girl and the other, somewhat more elaborate, about a boy; and together they are two of the most curious stories in the dream book. They are not even dreams, but there are other exceptions of this order in the book, and these two stories Freud expressly designs to illustrate the censorship. In fact, they enjoy a privileged status as uncommon inset narratives in Freud's theoretical chapter and the last such in his book, a mere four paragraphs from its close. Their position alone suggests the importance of the censorship to the theory, while it underlines the need for specific cases if the institution is to be comprehended. The story of a girl consulting Freud has no specific relations to other material in the book but is no less interesting on that account; the fourteen-year-old boy seems more like an old friend, who might be Freud himself.

The two stories, especially the first, are virtually self-contained. Freud prefaces them by writing, "Examples of every possible variety of how a thought can be withheld from consciousness or can force its way into consciousness under certain limitations are to be found within the framework of psychoneurotic phenomena; and they all point to the intimate and reciprocal relations between censorship and consciousness." Then he provides, in separate paragraphs, what he calls "a report of two such examples." Here is the first:

> I was called in to a consultation last year to examine an intelligent and unembarrassed-looking girl. She was most surprisingly dressed. For though as a rule a woman's clothes are carefully considered down to the last detail, she was wearing one of her stockings hanging down and two of the buttons on her blouse were undone. She complained of having pains in her leg and, without being asked, exposed her calf. But what she principally complained of was, to use her own words, that she had a feeling in her body as though there was something "stuck into it" which was "moving backwards and forwards" and was "shaking" her through and through: sometimes it made her whole body feel "stiff." My medical colleague, who was present at the examination, looked at me; he found no difficulty in understanding the meaning of her complaint. But what struck both of us as extraordinary was the fact that it meant nothing to the patient's mother—though she must often have found herself in the situation which her child was describing. The girl herself had no notion of the bearing of her remarks; for if she had, she would never have given voice to them. In this case it had been possible to hoodwink the censorship into allowing a phantasy which would normally have been kept in the preconscious to emerge into consciousness under the innocent disguise of making a complaint. (618)

Except for the prefatory sentence already quoted, then, this narrative stands by itself without any further interpretation than that offered by the telling. A thoughtful reader will immediately sense that the story omits some facts necessary to show "the intimate and reciprocal relations between censorship and consciousness" and admits others, brief as it is, that are gratuitous, such as the dress of the patient and the seeming negligence of the mother—features of the case singled out as "surprising" (befremdend) and "extraordinary" (merkwürdig). We can hope that the girl consulting Freud lingered and asked questions; the critic or historian approaching the passage is obliged to ask questions. What kind of narrative is this? How do the gratuitous details affect its reading? What *is* "the meaning of her complaint"? What specific conclusions about the censorship does this narrative support?

The ease with which Freud tells of the consultation and its persistence in successive editions of *The Interpretations of Dreams* argue that he does not perceive any problems with the kind of story it is: he calls it a "report" (Mitteilung). Yet for a clinical report the narrative is distinctly abbreviated as well as brief. The point being to illustrate the censorship, it may not matter that no disposition of the case is indicated. A more serious omission is the record of any inquiry into the state of the girl's knowledge of sexual intercourse. Unless her physicians learn this much, they could hardly be sure the body language she uses to elude the censorship is unconscious. Also, the tone of the narrative is not very clinical. A lay reader, I am the one to supply the words "sexual intercourse." Freud and his colleague communicate entirely with looks and innuendo. Then Freud continues to employ this innuendo with the readers of the dream book. The two doctors are as coy about sex as their patient is ignorant.

Despite the scene of consultation, "report" does not suit this story very well. The narrative is more nearly an anecdote, and a risible anecdote at that. Indeed, reading it as an obscene joke— subclass, doctor-patient joke—helps considerably in appreciating the cast of characters, the progress of the story, and the signals it sends forth. In *Jokes and Their Relation to the Unconscious*, Freud described such stories as refinements of "smut" (Zote), or the deliberate naming of sexual matters by men approaching women. "Smut is like an exposure of the sexually different person to whom it is directed," an exposure which both parties desire, but he actively and she passively. When the woman resists any physical approach, "the sexually exciting speech becomes an aim in itself"; and under conditions of social refinement, speech is also inhibited and disguised as a joke, even when the woman is not present. "The men save up this kind of entertainment, which originally presupposed the presence of the woman who was feeling ashamed, till they are 'alone together.' So that gradually, in place of the woman, the onlooker, now the listener, becomes the person to whom the smut is addressed" (1905a, 97–99).

By Freud's later account, then, this anecdote tells of the transformation of smut into a joke. The smut is spoken naively by the patient, however, and the seduction prevented by the clinical setting—or by the presence of the other doctor, who materializes in the story just when he is needed (we are not told he is there until after the girl speaks of something "stuck into it") and in harmony with Freud's own prescription for such jokes: "The ideal case of a resistance . . . on the woman's part occurs if another man is present at the same time—a third person—, for in that case an immediate surrender by the woman is out of the question" (1905a, 99). Thus we can easily imagine another version of the story, a doctor-patient joke of the show-me-where-it-hurts kind for which the punch line is something like "Doctor, I feel something stuck into me!" But that joke would impart at least half of the ribaldry to the doctor, and what Freud treats as obscene is likely to be more self-protective, as in the present anecdote. "Generally speaking," he will write, "a tendentious joke calls for three people: in addition to the one who makes the joke, there must be a second who is taken as the object of the hostile or sexual aggressiveness, and a third in whom the joke's aim of producing pleasure is fulfilled." (All this, by the way, is brilliant literary analysis.) The third person in the present instance is the medical colleague, whisked into view by the narrative in order to raise an eyebrow with Freud. Moreover, "when the first person finds his libidinal impulse inhibited by the woman, he develops a hostile trend against that second person and calls on the originally interfering third person as his ally" (1905a, 100). Some hostility would seem apparent in Freud's remarking of the patient's dress as well as the supposed negligence of the mother, who materializes in the story just as neatly as the colleague.

Readers of the dream book encounter this doctor-patient joke as third persons also, as the medical colleague becomes multiplied over and over again in print. Through such dramatic irony readers become superior to patients—they are granted privileges—in all the narratives emerging from Freud's consulting room in all his writings. This story has patently been assembled for the pleasure and instruction of readers—especially instruction, of course, since it is to serve as an example of the censorship. For this reader, its penury as a report and similarity to a joke cast its provenance completely in doubt. The textual history of the passage is in no way reassuring, since the words the patient supposedly spoke were not originally distinguished from the rest. The early editions of *Die Traumdeutung* print the central portion in spaced type, as follows: "Ihre Hauptklage aber lautet wörtlich: S i e   h a t   e i n   G e f ü h l   i m   L e i b e ,   a l s   o b   e t w a s   d a r i n   s t e c k e n   w ü r d e ,   w a s   s i c h   h i n   u n d   h e r   b e w e g t   u n d   s i e   d u r c h   u n d   d u r c h   e r s c h ü t t e r t .   M a n c h m a l   w i r d   i h r   d a b e i   d e r   g a n z e   L e i b   w i e   *s t e i f.*" (I quote from the fourth edition, Leipzig and Vienna, 1914, p. 478; the first edition is the same except for differences in orthography.) Thus the punch line was originally delivered entirely in the indirect style with added emphasis on the last word, *stiff.* Perhaps because Freud suggested that he was presenting the patient's complaint word for word (wörtlich), the editors of the *Gesammelte Werke* selected for emphasis only those words she could actually have uttered. The standard edition then prints these few words within quotation marks, which signal to the reader that they were hers.

The anecdote, in short, seems crafted to suit Freud's purposes, though its original styling did not claim quite so much as the present version. But even make-believe may serve serious purposes, as fictitious cases serve law professors' purposes, or verbal problems algebra teachers'. So we should inquire further: how do the gratuitous details affect one's reading of the story? The facts represented by this narrative begin immediately and extend just beyond the climax—the sentence that expresses the patient's symptoms—to the sentence introducing the colleague to the scene: that the colleague "looked at me" and Freud's interpretation of the look still refer to the scene of consultation. Then, with the sentence about the mother, the anecdote moves into post-factum conjecture, a sentence of deductive reasoning, and a sentence allegorizing the story as one of censorship. Some of the observed facts are the patient's unembarrassed manner (unbefangen blickenden), the state of her clothing, her complaint about her leg and exposure of her calf, and the reported sensation of something "stuck into" her. These details provide a degree of suspense and prepare for a sexual revelation, so that when the climax is reached we are not taken completely by surprise. A perfect gloss is available, again from Freud's discussion of obscene jokes: "In women the inclination to passive exhibitionism is almost invariably buried under the imposing reactive function of sexual modesty, but not without a loophole being left for it in relation to clothes" (1905a, 98). As for exposing her calf without being asked, that action definitely prepares us for something risqué. Freud also mentions her intelligence, which may make her lack of embarrassment seem less innocent. Possibly her description owes something to the late nineteenth-century lore about Viennese girls that Sander L. Gilman has documented (1985, 39–58). She may be so driven by the inclination of women that she is putting the doctors on.

The details supplied for preparation and suspense have almost been too well told. That is one reason for Freud's assertion, in the penultimate sentence, that she "had no notion of the bearing of her remarks," a seeming a priori assertion for which no fact is in evidence. Since she did not, or rather could not, know what she was talking about, what she said must have eluded the censorship. Other constraints on the possible significance of the story are simply taken for granted. Even the age of the patient need not be mentioned directly, because it is sounded by special alarms. She is a girl in the process of becoming a woman, who ought to abide (it is certainly surprising that she does not abide) the rule for woman's clothes, which "are carefully considered down to the last detail." The rules apply especially when the young woman is still a virgin, and—after the revelation of the patient's symptoms subsides—alarm is accordingly expressed at her upbringing. It is not even clear—no evidence is presented— that the doctors have exchanged any words with the mother. But the latter is implicitly to blame for her daughter's appearance and manners and surprisingly remiss in not checking the recitation of her symptoms. Somehow this is a mother's problem and should not have required a doctor (perhaps the reason we hear nothing of the disposition of the case). For her negligence and because some hostility has been aroused, the doctors cap the joke by mentally sticking it to the mother, who "must often have found herself in the situation her child was describing." For sheer objectivity in regarding the opposite sex, you cannot go much further than that.

Behind the raised eyebrows of the physicians, the relevant clinical question of what the patient knows about sex is abandoned for what the mother has certainly felt. And what the mother knows, or ought to know, is the rule that women never openly show that they know. The look exchanged by the two doctors and Freud's innuendo imply that they, at any rate, uphold the polite secret of human sexuality, which the mother has failed to teach her child. This is not to contend that Freud should have been more open but to point to the original and present location of the censorship in the story. The girl, through fault in upbringing, evidently does not understand the rules that Freud and his imagined readers just as evidently subscribe to. As we can begin to see, these rules *are* the censorship, for Freud has no other grounds for inferring what thoughts might or might not be blocked from consciousness. And the seemingly unnecessary details of the anecdote—the girl's unembarrassed manner and careless dress, her showing of her calf without being asked—belong to the same story—or history—of social censorship. They mark an ignorance of the rules, not the inward unconsciousness of sex.

Then what is "the meaning of her complaint"? Initially this is an indirection, a periphrasis, a cover for sexual intercourse—which in the original is still more indirect (er findet die Klage nicht missverständlich). Once the girl has so remarkably put her sensations into words, in each remaining sentence of the anecdote the narrator covers for her: "the meaning of her complaint," "the situation which her child was describing," "the bearing of her remarks," and "a phantasy." Each time he performs this operation with words—a form of social censorship—he accedes to the prudishness so ingrained in the language of the culture that it passes unnoticed; and each time, he claims the superior knowledge of sexuality of one who need not descend to details. There is nothing personal in this prudishness and superiority, which are of course shared with the reader in the dramatic irony of the anecdote. If Freud or his colleague had stopped to explain sex to their patient then and there, his readers would be let down completely.

"The meaning of her complaint" also stands for the principal fact in issue, the censorship. The patient's complaint—the fact that she made such a complaint—is supposed to show "relations between censorship and consciousness." So in translation the periphrasis serves also as Freud's conclusion, that which was to be demonstrated. Still, the very knowingness conveyed through all the remaining sentences has to be a little uncertain, because at the last moment the narrator tries to recast the whole argument as a bit of deductive reasoning: "The girl herself had no notion of the bearing of her remarks; for if she had, she would never have given voice to them." Buried within the conditional that serves as this syllogism's major premise is the most confident assertion of causality in the entire narrative. The conditional itself—if she had known, she would not have spoken—speaks of social conditions once more: those conditions impending on such girls at this time and place in history. Two things can be said about the tactic of burying this induction beneath the rhetoric of deduction. First, Freud discards in this sentence the evidence he himself has offered that this particular girl is different from others. Second, though we may grant the induction (he being a better judge of Viennese girls in 1900 than we are), in the sentence that follows he simply transposes this description of social censorship, having to do with what one may or may not say before others, to a description (equally general) of internal censorship, having to do with what thoughts can or cannot pass from the preconscious to consciousness. And whereas his tactic has excluded the possibility of any exception to the social censorship, he immediately pleads an exception for the internal censorship: "In this case it had been possible to hoodwink [abzublenden] the censorship."

The agency of this exception, the hoodwinker or cause of hoodwinking, is never specified in the narrative; no more is the source of the "phantasy" of being invaded by the male. Readers have to understand, from other contexts in *The Interpretation of Dreams*, that an unconscious wish is supposed to be operative here. Freud neither spells out the unembarrassed girl's motive nor reflects that mostly girls do not complain of such symptoms. Yet despite the kind of story this is, its doubtful provenance, gratuitous details, and indirections, it clearly is consistent with two basic observations about the censorship. A temerarious historian would claim that *all* Freud's stories of the censorship corroborate these two points.

First, Freud's metaphor is basically drawn from social and not political censorship. The careless girl, her mother, and her doctors are not concerned with politics as such, not even the futile branch of politics that is literary censorship. In Freud's dreams that have political implications, the censored thoughts similarly relate to unacceptable behavior or attitudes generally and reflect kinds of dissimulation evident in many forms of politeness. The prohibited action or attitude need not even be immoral but merely of a kind not generally approved. Freud keeps in view at least as wide a range of social constraints as John Stuart Mill fending against "a hostile and dreaded censorship" in *On Liberty* (1859, 264). The parameters of constraint attributed to the internal censorship are the same as those of the social censorship. Throughout *The Interpretation of Dreams* Freud assumes that in the mind lie concealed precisely the sorts of thoughts that are concealed "where two persons are concerned" in social life. He in each case takes for granted the specific impropriety or threat of disclosure and relies equally upon his readers' acquaintance with social rules. He depends implicitly on prior agreement as to what is scandalous. He has no other grounds for inferring what the internal censorship would censor. The anecdote of the careless girl illustrates these assumptions perfectly. When Freud succumbs to the deductive rhetoric of "the girl herself had no notion of the bearing of her remarks; for if she had, she would never have given voice to them," he thereby exposes the limitations of the analogy to social life with which he began and which he has institutionalized as "the censorship." Though this agency may be thanked for the glimpses it may give of thoughts that are unexpressed, what it allows or disallows can only be gauged by some previous measure of social discourse.

Second, the censorship is like nothing so much as an act of failed and clumsy repression. From the beginning Freud closely associated censorship with repression (1896, 182–85), but his theory in the long run required two sorts of checks on the unconscious: one steady and predictable and the other a far less certain "agency." Because "unconscious wishes are always on the alert, ready at any time to find their way to expression" (553), the checks required should be weighty, and at first the statist connotations of censorship promised to supply the weight. Also, Freud's theory is not of instincts but of unconscious ideas, and a censorship is the kind of institution equipped to stop or regulate the flow of information across borders. Unfortunately for this institution, however, unconscious ideas represent instincts, as he later put it (1915, 177), and the hapless censorship finds itself in an age-old battle of prudence with passion that the humorist in Freud cannot resist. In the anecdote of the careless girl, his joke leaps to the ribald conclusion that she is eager to engage in sex, and only the poor censorship, in the business of sifting ideas, supports a conclusion that she might only want to know about sex. In his section on dreams of exhibiting oneself, Freud dutifully writes, "The unconscious purpose requires the exhibiting to proceed; the censorship demands that it shall be stopped" (246). But in practice the censorship's demands are made a mockery of; in fact the opposition of unconscious purpose and the censorship is seldom so squared off as that formulation suggests. Far from being able to boast of the many ingenious ways it suppresses thought, this censorship usually leaks the thought without delay.

As Freud has admitted right along, the censorship enjoys "the privilege . . . of permitting thoughts to enter consciousness" (144). The censorship, we may think, thus forces wishes to show their stuff. "Under the pressure of the censorship," Freud suggests in his book on jokes, "any sort of connection is good enough to serve as a substitute by allusion, and displacement is allowed from any element to any other" (1905a, 172). If any sort of connection is good enough to get by, little wonder the censorship is bested in story after story and never succeeds in its primary effort of suppressing a wish. As the consultation with the sloppily dressed girl shows, the censorship is mainly there to be hoodwinked, bamboozled, or thrust aside. In the ample frame story of *The Interpretation of Dreams*, the censorship is both ringmaster and clown, first creating and then bungling the act. As in the circus, an alluring young woman can easily fool the censor.

Having reviewed the inherently social nature of Freud's claims for the censorship, I wish to draw back from the story of the girl to a vantage that includes the story of the fourteen-year-old boy, which follows it without a break and is positively the last narrative in the great book. Though the censorship is once again cast in its telltale role of betraying rather than protecting secrets, the story of the boy is markedly different from that of the girl. Freud this time primes the censorship for failure, by inviting the patient to describe whatever pictures come to him when he closes his eyes. The boy's tic, vomiting, and headaches are entirely uninteresting symptoms compared to the fascinating complaint of the girl, but the pictures permit readers to see into his subjective life—a perspective utterly closed to us in the case of the girl. This subjectivity of the male is then substantiated by allusions to some of Freud's own favorite stories from mythology, including the apparently mistaken notion that Zeus castrated his father:

> Here is another example. A fourteen-year-old boy came to me for psycho-analytic treatment suffering from *tic convulsif,* hysterical vomiting, headaches, etc. I began the treatment by assuring him that if he shut his eyes he would see pictures or have ideas, which he was then to communicate to me. He replied in pictures. His last impression before coming to me was revived visually in his memory. He had been playing at draughts with his uncle and saw the board in front of him. He thought of various positions, favourable or unfavourable, and of moves that one must not make. He then saw a dagger lying on the board—an object that belonged to his father but which his imagination placed on the board. Then there was a sickle lying on the board and next a scythe. And there now appeared a picture of an old peasant mowing the grass in front of the patient's distant home with a scythe. After a few days I discovered the meaning of this series of pictures. The boy had been upset by an unhappy family situation. He had a father who was a hard man, liable to fits of rage, who had been unhappily married to the patient's mother, and whose educational methods had consisted of threats. His father had been divorced from his mother, a tender and affectionate woman, had married again and had one day brought a young woman home with him who was to be the boy's new mother. It was during the first few days after this that the

fourteen-year-old boy's illness had come on. His suppressed rage against his father was what had constructed this series of pictures with their understandable allusions. The material for them was provided by a recollection from mythology. The sickle was the one with which Zeus castrated his father; the scythe and the picture of the old peasant represented Kronos, the violent old man who devoured his children and on whom Zeus took such unfilial vengeance. His father's marriage gave the boy an opportunity of repaying the reproaches and threats which he had heard from his father long before because he had played with his genitals. (Cf. the playing at draughts; the forbidden moves; the dagger which could be used to kill.) In this case long-repressed memories and derivatives from them which had remained unconscious slipped into consciousness by a roundabout path in the form of apparently meaningless pictures. (618–19)

As in the story of the careless girl, this story of a naughty boy eventuates in assertion rather than proof of any operation of the censorship. The censorship, indeed, is the one element of the patient's mental life that has failed in its office and turned up at the consultation like a spy or blackmailer with pictures that threaten to reveal all. Nevertheless, in this story the rest of the details are far from gratuitous, since no plot attaches to these pictures until the family history and classical mythology are set forth, coupled with the requisite identifications ("His suppressed rage against his father *was* what had constructed the pictures," "The sickle *was* the one with which Zeus castrated his father," etc.). This "report" about the censorship simply would not be comprehensible without these social and literary components, which depend implicitly on what sorts of situations might constitute an "opportunity" (Gelegenheit) for childish vengeance in the first instance and classical precedent for all that weaponry in the second. Freud fails to state how he knows about the boy's rage or his acquaintance with mythology, but this time the details he does provide tell boldly of conflict and mental life.

Moreover, the boy's guilty thoughts and behavior cannot seriously be condemned as immoral. The carefully retarded narrative of masturbation, saved for the next to last sentence, provides a weak climax to the series of revelations—more affecting for readers in 1900 than today perhaps. It is the one aspect of the patient's actual behavior that is mentioned, and unlike the careless girl's behavior this act is assumed to have a history of some significance, even if Freud merely intuited it (he does not say how he heard of it). Such psychological, as opposed to moral, guilt is one of the principal lessons of Freud's teaching, and one of the abiding pleasures of his stories is their dwelling typically on harmless sexuality and harmless aggression—harmless as far as other people are concerned. "So it makes no difference whether one kills one's father or not—one gets a feeling of guilt in either case!" he imagines readers protesting thirty years later, before explaining once more the difference between psychological guilt (Schuldgefühl) and remorse (Reue) (1930, 131–32). By then Freud had moving and deeply pessimistic things to say about the cumulative force of aggression within civilization itself, but his individual narratives of psychological guilt characteristically bring pleasure and relief: once one copes with such feelings by confession to the doctor, one can savor the boldness of unconscious aggression. This form of romance verges on fairy tale rather than an obscene joke and is distinctly subjective rather than objective in its appeal.

The idiosyncratic telling of the myth of Zeus and Kronos, to which Freud called attention soon after writing this passage, argues that his tale of a naughty boy is quite as blatantly a fiction as the joke about a careless girl. As we have seen, even as the dream book was appearing in print, Freud began to formulate the idea that the errors in it were due not to carelessness but to unconscious wishes (1985, 385). Writing a year later in *The Psychopathology of Everyday Life*, he admitted he was possibly wrong about Zeus emasculating Kronos, and wrong in two places in the previous book: namely, in generalizing earlier about dreams of the death of close relatives and in the present story. In his "Errors" chapter he did not bother to distinguish the two very different contexts but baldly stated, "The errors are derivatives of repressed thoughts connected with my dead father" (1901, 219). From this casual admission it would seem that the fourteen-year-old's knowledge of mythology is Freud's knowledge, and the story so thoroughly subjective in its telling as to erase the distinction between the boy and its narrator. Two just visible seams in the narrative ("After a few days I discovered the meaning of this series of pictures" and, further on, "The material for them was provided by a recollection from mythology") are noticeably noncommittal as to the site of the discovery or the recollection. In autobiographical passages, this order of deception has come to be taken for granted by Freudians, but if this story is autobiographical, it leaves behind a bad taste in such fanciful details as the parents' divorce or the character of the father.

It so happens that the stories of the girl and the boy have a connection that antedates by at least five years their appearance side by side in the theoretical chapter of *The Interpretation of Dreams*: a connection that belies any possibility that the consultation with the girl occurred "last year" and leads one to suspect that a good many other narratives in the book were made to order. The malleability of such fictions is apparent, for as it happens, an outline of these two was first roughed in by a paragraph in Freud's theoretical contribution to *Studies on Hysteria* that began with a similar offer of two examples:

> I will give one or two examples of the way in which a censoring of this kind operates when pathogenic recollections first emerge. For instance, the patient sees the upper part of a woman's body with the dress not properly fastened—out of carelessness, it seems. It is not until much later that he fits a head to this torso and thus reveals a particular person and his relation to her. Or he brings up a reminiscence from his childhood of two boys. What they look like is quite obscure to him, but they are said to have been guilty of some misdeed. It is not until many months later and

after analysis has made great advances that he sees this reminiscence once more and recognizes himself in one of the
children and his brother in the other. (1895, 282)

Here "the patient" instead of the two doctors pictures the
woman *and* reminisces about his childhood. But as in the later
version of two stories, we meet with a woman with her dress
unbuttoned and a guilty boy, while the focus on what the patient "sees" anticipates the method of calling up pictures in the
case of the fourteen-year-old boy. Though the later boy has no
brother to be reckoned with as here, Freud would casually refer
to *his* brother when confessing his error about Zeus in *The Psychopathology of Everyday Life*. What Freud calls censoring in this
earlier passage is closer to what he generally refers to as resistance, but the anticipation of that metaphor in connection with
the two examples also links them to the stories he chose to place
last in the dream book. I cannot even guess at the personal significance of the stories, if any, but they are clearly fiction as
presented.

The penultimate example of censorship in *The Interpretation
of Dreams* is a smartly objective story of a shameless girl, and the
ultimate example a complacently subjective story of a guilty boy:
positioned together as they are, the stories would seem to echo
a respectable homiletics rather than science or autobiography. In
an early eighteenth-century discussion of "how necessary an In-
gredient Shame is to make us sociable" and of repression as the
provisional "stifling of our Appetites," the physician Bernard
Mandeville gave it as his opinion that different expectations
about the behavior of the two sexes were mainly due to early
education. "*Miss* is scarcely three Years old," Mandeville wrote,
"but she is spoke to every Day to hide her Leg, and rebuk'd in
good Earnest if she shews it; while *Little Master* at the same *Age*
is bid to take up his Coats, and piss like a Man" (1714, 68, 72).
It can be shown, it seems to me, that Freud continues to teach
this wisdom two centuries later, from a perspective not unlike
Mandeville's. He may be as knowing as he pleases—and this

knowingness pleases his readers—but for the behavior proper to each sex he defers entirely to the culture. Women he believes are inclined to fall, while the dream book as it applies to men is in many of its pages a homily on modest ambition. His last two stories of the censorship depend on preconceptions about boys and girls and go on to make their own contribution to the education of modern patriarchs. Roughly, boys are naughty when they masturbate instead of courting or seducing girls, and girls behave shamelessly when they expect to be courted or seduced before they are asked.

Even the shadowy version of the two stories in *Studies on Hysteria* registers an agreed gender difference, since in that version the woman has no being whatever except in the subjective gaze of the man. In the more elaborate version of the dream book, the young woman is presented through externals only and treated as a creature of shame. She may dress either correctly or incorrectly, behave in an accepted or unacceptable manner, look upon the world with or without embarrassment, but unfortunately she loses on all these counts. Shame would have prevented her from saying anything at all, if she had known what she was talking about. Freud patently subscribes to the convention against showing the leg that Mandeville remarks; elsewhere it is evident that he subscribes to a general doctrine that women are more subject to shame than men. For all his self-criticism and close observation of his own surreptitious thoughts, it was apparently as unthinkable for him to feel ashamed as it was for any bourgeois who set up as paterfamilias. In one extraordinary jotting among the papers he sent to Fliess in 1896, he remarks, "Where there is no shame (as in a male person), or where no morality comes about (as in the lower classes of society), or where disgust is blunted by the conditions of life (as in the country), there too no repression and therefore no neurosis will result from sexual stimulation in infancy" (1950, 222). One need read only as far as the first parenthesis. Toward the end of his life Freud would call shame "a feminine characteristic *par excellence*" and attribute it to women's "concealment of genital deficiency" (1933, 132), thereby addressing one more time the principle of pissing like a man. In viewing women so, he deprives them of any interiority save that which he jocularly attributes to the careless girl's mother—the sensation of a penis moving around inside her.

Now shame *is* the internalized estimate of how others whom we care about regard us. But Freud never seems to have given much thought to it or to have been terrified or sickened by shame, as most men and women have occasion to be, especially when they are young but when they are older as well. For him interiority is importantly evidenced by guilt and the management of guilt, because that is essentially the teaching of Judeo-Christian culture about inner being. Hence the naughty fourteen-year-old boy is scarcely ashamed of anything but possesses an inner life and value incommensurate with a girl's, and Freud cannot properly be laughed at for carelessness about Zeus and Kronos, because he was deeply engaged at that time in battling the ghost of his father. Of course Freud's prejudices tend to take over, as they do with most people trying to categorize others in some way different from themselves. Roy Schafer acutely observes that "Freud seemed to know that he was not making empirical assertions about how women really and necessarily are." But of Freud's attitude to mothers in particular, Schafer justly adds, "In his writings he showed virtually no sustained interest in their subjective experience—except for their negative feelings about their own femininity and worth and their compensatory cravings to be loved and impregnated, especially with sons" (1977, 354, 357). Nothing is very surprising in all of this: the last two stories in the dream book reflect the social perceptions of their time as much as the idiosyncrasies of their author.

The censorship was bound to be a rather odd institution, since it both gets in the way and gives an analyst something to do. The practice of psychoanalysis generally assumes that a person whose unconscious wishes are censored is the better for discovering what they are, and this assumption may already be implicit in *The Interpretation of Dreams*: it is instructive as well as pleasing to know the secret meaning of our dreams. At the same time, the alliance of the censorship with repression argues that it ought to have a stabilizing function in mental life. Freud never took the position that we could exist as familial and social beings without repression. At moments it does seem that the censorship—the bad guy in the romance of his book—has a respected duty to perform, though such moments may be rare. At one point in the theoretical chapter, Freud writes, "Thus the censorship between the *Ucs.* and the *Pcs.*, the assumption of whose existence is positively forced upon us by dreams, deserves to be recognized and respected as the watchman of our mental health" (Wächter unserer geistigen Gesundheit) (567). But how strangely this watchman is shown off to us in the book of dreams! He can only be a comic figure for repression, since he takes tumble after tumble before the appreciative eyes of the book's readers.

In ancient times the censor was a Roman stalwart, and perhaps he comes in for a beating from Freud on that account. To be sure, in the dream book the character persistently appears as the censorship (die Zensur) rather than the censor (der Zensor), but in German the words scarcely differ except in gender. As in the circus clowns often dress as women, so in psychoanalysis the censorship may be a personification dressed as an institution. Certainly its performances are at once agile and prone to failure. Its failures are part of the act, which can be repeated over and over again with undeniable success. This act delights us—and has accustomed us to believe in the power of unconscious motives that predictably overcome the censorship. But in each instance the supposed censorship has been inferred from social life, and as such it does not provide evidence of the dream work or of the unconscious. When Freud's accounts of censorship in the dream book ring true, that is because we are acquainted with some "similar distortion . . . in social life" (141). The success of the censorship testifies to the writer's feel for comedy, his sense of an audience, and his skillful representation of the rules, embarrassments, and concealments of life as he understood it.

CHAPTER FIVE

"The Only Villain among the Crowd of Noble Characters"

THE LAST TWO STORIES embedded in the dream book certainly pose something of a puzzle. They are not about dreams, they feature the censorship, and—as I trust I have shown—they are spurious. The first is blatantly anecdotal, a joke on women clearly meant to amuse; the second is part fiction, part autobiography, and finally neither one nor the other. Yet the pair of stories, supposedly about actual consultations involving a careless girl and naughty boy, offer the last evidence in the book for Freud's theory as well as the last bit of humor. Except for the humor, they seem so unwarranted that one wonders why they are there at all. If *The Interpretation of Dreams* came down to us from an ancient manuscript, or in a folio like that of Shakespeare's plays, scholars would long ago have written off the stories as scribal interpolations. But this is a modern book, revised through a good many editions by Freud himself.

So we are bound to believe that Careless and Naughty were meant to play roles intrinsic to the closing down of Freud's masterpiece, if only—after their fashion—in the homiletics. So dubious or mischievous a pair do help enliven a very sober theoretical chapter. I have suggested that the censorship itself functions as a clown in the romance of the book and produces a generally happy effect. There may have been something of the grave clown in the author who created all the pleasing effects, and this impression is reinforced by the four paragraphs that complete the

dream book. After so long a work with so many centers of interest, the final paragraphs may be discounted by some readers. For
one thing, they are about a problem that Freud raises only to
dismiss: "the *practical* value of this study" and more especially
"the ethical significance of suppressed wishes" (619–20). As with
the two stories of censorship, a question arises as to why this
argument should appear at the very end. Freud's final paragraphs
are as lighthearted as his anecdotes and still more puzzling, for
if taken seriously they would sidetrack his whole program for
mapping the unconscious, by displaying the claims made about
repression and censorship as arbitrary. I find the ending of the
book both serious and not serious, embarrassed and confiding.

In context, it is quite clear that by "practical" Freud means
not clinical uses but ethical uses: whether dreams should be consulted as evidence of character or moral intentions. He contrasts
practical and theoretical in a philosophical sense:

> Thus I would look for the *theoretical* value of the study of
> dreams in the contributions it makes to psychological
> knowledge and in the preliminary light it throws on the
> problems of the psychoneuroses. . . . But what of the *practi-*
> *cal* value of this study—I hear the question raised—as a
> means towards an understanding of the mind, towards a
> revelation of the hidden characteristics of individual men?
> Have not the unconscious impulses brought out by dreams
> the importance of real forces in mental life? Is the ethical
> significance of suppressed wishes to be made light of—
> wishes which, just as they lead to dreams, may someday
> lead to other things? (619–20)

Freud at first replies that he does not "feel justified in answering
these questions" without more consideration, but on the whole
he would exempt dreams from any ethical test or significance. "I
think it is best, therefore, to acquit dreams," he writes. "Whether
we are to attribute *reality* to unconscious wishes, I cannot say."
In later editions, perhaps judging that this statement was incon-
clusive, he added, "If we look at unconscious wishes reduced to

their most fundamental and truest shape, we shall have to conclude, no doubt, that *psychical* reality is a particular form of existence not to be confused with *material* reality" (620; the editor
observes that in 1919 *material* was substituted for *factual*). Early
and late, Freud seems to have been groping a little in composing
these sentences, but their purport is agreeable enough. The
phrase most often quoted, after all, is his decision "to acquit
dreams."

Where he flounders, it may be, is in the following penultimate
paragraph, an attempt to shore up this acquittal by observing
that dreams are not needed to judge character.

> Actions and consciously expressed opinions are as a rule
> enough for practical purposes in judging men's characters.
> Actions deserve to be considered first and foremost; for
> many impulses which force their way through to conscious-
> ness are even then brought to nothing by the real forces of
> mental life before they can mature into deeds. In fact, such
> impulses often meet with no psychical obstacles to their
> progress, for the very reason that the unconscious is certain
> that they will be stopped at some other stage. It is in any
> case instructive to get to know the much trampled soil from
> which our virtues proudly spring. Very rarely does the com-
> plexity of a human character, driven hither and thither by
> dynamic forces, submit to a choice between simple alterna-
> tives, as our antiquated morality would have us believe.
> (621)

(The last paragraph of all—a pleasant conceit that dreams do tell
of the future, since the notion of wishes fulfilled points in that
direction—is harmless enough. It may even speak, as in the epi-
logue of an old book, of the hoped-for success of this one.)

Freud is heading for trouble in this paragraph before he res-
cues himself with the hit on our supposed virtues and antiquated
morality. The middle sentence, for one thing, imagines an unconscious that calculates on its thoughts being checked. This
new state of affairs contravenes in three respects the theory that

he has constructed: it implies that the unconscious in other cases (when it is *not* certain its impulses will be checked at some other stage) engages in checking itself; it imputes knowledge of the whole personality and even a degree of moral judgment to the unconscious; and it discards the censorship in favor of some fully conscious means of self-control at that other stage. These un-happy contradictions are laid in store by the previous sentence, with its admission that "many impulses which force their way through to consciousness are even then brought to nothing by the real forces of mental life" (durch reale Mächte des Seelen-lebens). Freud is explaining, at this late moment in the book, why unconscious wishes need not be judged in an ethical light or serve as evidence of character. In doing so, however, he not only raises the question of the possible insignificance of the dreams he has studied but suddenly exposes the mechanisms of repression and censorship as completely arbitrary. And what are the "real forces" faced up to in this next-to-last paragraph? Have not the forces of mental life been the object of study? It seems that the interrelations of all such forces ought to have been taken up much earlier.

Readers are suddenly reminded how slight and mischievous the latent dream thoughts uncovered by analysis have been in comparison with immoral and even criminal plans that may be entertained by anyone and sometimes acted upon. Even thoughts such as these need not be taken as evidence of charac-ter, since according to the first sentence "actions and consciously expressed opinions" are more to the point. Yet by the same ac-count consciousness can be chock-full of truly wicked ideas, so what grounds are there for asserting that similarly wicked or (as in most instances adduced by the dream book) far less wicked ideas are repressed or censored? The dangers risked here are not only to Freud's theory but to the case-by-case argument by story-telling that provides the pleasing matter of *The Interpretation of Dreams*. The jolt administered by the real forces of mental life rocks the dream analyses all the way back to that of Irma's injec-tion. How could Freud help knowing that he would like to place the blame for "Irma" on his colleagues or to be rid of this or that rival in the laboratory? And since he did know—as this paragraph at the end pretty well admits, we all know such things—why would not the many stories in the book about latent dream thoughts have been formed by his conscious knowledge? One of the reasons the stories are pleasurable is that they are indeed based on a knowledge of motives shared with readers and confirmed in some way; Freud has used the principles of narrative suspense to good purpose. But the turn taken by this argument at the end is devastating for his claims about the unconscious, unless we are willing to be diverted by the *sententiae* that he brings in to close the paragraph. It is as if the royal road to the unconscious were being mined by a practical joker—or the court jester.

Significantly, Freud returned to the question of ethics and dreaming many years later, in some pages evidently intended to be added to the end of *The Interpretation of Dreams*. Answers to this question may differ, of course, as to whether the manifest or latent content of a dream is meant. He dismisses the need to worry about the former: "We know now that the manifest content is a deception, a *façade*. It is not worth while to submit it to an ethical examination or to take its breaches of morality any more seriously than its breaches of logic or mathematics" (1925, 131). But this reflection gives rise to another about censorship. How is it that the manifest dream contains any immoral thoughts, if the censorship has done its job? As we have seen, this is a question that teased the author now and then in the initial writing of the book but never surfaced quite so frankly as here, in the reconsideration of ethical significance. Twenty-five years on, Freud accurately states one of the objections I have just raised to the end of his book. "How can it happen, then," he asks, "that this censorship, which makes difficulties over more trivial things, breaks down so completely over these manifestly immoral dreams?" Unfortunately, he doesn't really have an answer; or in his words, "The answer is not easy to come by and may perhaps not seem completely satisfying." Some dreams are uncensored "because they do not tell the truth," but others because they tell horrible truths without need of interpretation. Finally he overrides the objection: it simply is not of much interest, if one reflects that "the majority of dreams . . . are revealed, when the distortions of the censorship have been undone, as the fulfilments of immoral—egoistic, sadistic, perverse or incestuous—wishful impulses." So it really makes little difference whether the censorship is in good operating form or not! One way or another, most dreams are the fulfillments of immoral wishes. As if to second my belief that this whole psychology is conditioned by imaginative constructions of the late nineteenth century—above all, that secrets must be suspected everywhere— he drives home his point with a simile that might literally appear in a detective or sensation novel: "As in the world of waking life, these masked criminals"—the aforesaid wishes—"are far commoner than those with their vizors raised" (1925, 131–32).

The later elaboration draws out the denouement of *The Interpretation of Dreams* without altering its tactics or rendering it straightforward. The critical questions of Freud's placement of the two dubious stories of censorship and his paragraphs on the "practical" value of the book remain. The ending of his masterpiece is a little embarrassed, a little whimsical, a little confused, a little defiant. Freud at once cedes the existence of problems and passes them off lightly. The censorship is a bit of a joke, and dreams are of no practical significance. By these daring suggestions, and trusting to his readers' identification with him, the author forestalls criticism by rather unusual means. As famous clowns discovered they could get their audiences to identify with increasingly subtle antics in the circus, Freud in effect dares his readers to go along as he lightheartedly pulls examples of the censorship from his sleeve and indirectly reveals that he has made no firm distinction between conscious and unconscious motives after all. "And the value of dreams for giving us knowledge of the future?" (621). Even the little play on this impossibility is part of the performance. If the act succeeds—as it demonstrably does for great numbers of readers—a certain complicity is joined between the audience and the performer, who exits on the run with an exaggerated bow, a wave, and a wink. Criticism has been forestalled because Freud himself has once again raised the right questions, but more subtly and joyfully because his readers have in one degree or another identified with the effrontery and risks of the performance. As in most really moving clown acts, a profound embarrassment binds the audience and performer.

Though I am in danger of travestying the end of the dream book, I hope the point will not be lost (often a second clown is called for to put the first in his place). Freud comes close to admitting the appeal of his work as a performance earlier when he harps on the dangers to the author of indiscretion in telling of his dreams: those many protests and asides, after all, serve as reminders that he is willing to play the part. More than many writers he depends on repeated, persistent appeals for the reader's sympathy. Some direct appeals he delivers with the sobering modesty of scientific work, but modesty itself is a very delicate performance:

> In venturing on an attempt to penetrate more deeply the psychology of dream-processes, I have set myself a hard task, and one to which my powers of exposition are scarcely equal. Elements in this complicated whole which are in fact simultaneous can only be represented successively in my description of them, while, in putting forward each point, I must avoid appearing to anticipate the grounds on which it is based: difficulties such as these it is beyond my strength to master. . . . I am conscious of all the trouble in which my readers are thus involved, but can see no means of avoiding it. (588)

Modesty seems rational only because it is so conventional. In practice the modest worker contradicts himself by putting himself forward even as he depreciates his own powers. Habituated as we are to this elaborate restraint, no wonder we laugh aloud at clowning in the circus that both exaggerates abilities and culminates in utter failure. Every modest thought is an effort, strenuous of its kind, to mediate an irony of opposite appearances. I feel justified in classifying Freud's special modesty with that of professional performers because it generally exceeds the conventional mode illustrated above. His favorite posture, in *The Interpretation of Dreams* and later writings, is that of a maligned or unappreciated innovator in the back halls of science, whose ironical opposite is the acclaimed hero of the amphitheaters. Freud plays that part with an exaggerated irony that I am not alone in finding singularly appealing.

Let me instance one pronouncement on a favorite theme, from his second go-round with *Non vixit*. That dream and its interpretation register Freud's wish to survive even at mortal cost to his colleague P. in the laboratory. Memorably, he illustrates the relationship with a joke about a married couple. "I was delighted to survive, and I gave expression to my delight with all the naïve egoism shown in the anecdote of the married couple one of whom said to the other: 'If one of us dies, I shall move to Paris.' So obvious was it to me that I should not be the one to die." At this point (a telling point, reminiscent of David Copperfield's plight with his child-wife Dora), Freud reverts in a new paragraph to the theme of how embarrassing it is to unveil the meaning of his dreams: "It cannot be denied that to interpret and report one's dreams demands a high degree of self-discipline. One is bound to emerge as the only villain among the crowd of noble characters who share one's life" (485). In this instance the two sentences convey first the sententious complaint and then a mischievous irony. Freud is by no means a villain (Bösewicht), let alone a wrongdoer; and the crowd of so-called nobles includes his colleague P., who according to the story harbors wishes no less murderous than Freud's. The choice of "villain" is literary to start with, not at all equivalent in modern times to "criminal," whether in English or in German. "Who calls me villain?" Hamlet demands of himself, and Freud similarly indulges in a bit of playacting. The effect pleases and at the same time forestalls criticism, both in that it is pleasing and because any reader who enjoys the irony of opposites has already partially exculpated Freud. Irony exacts complicity; complicity makes allies, binds readers to him.

But let me instance another occasion on which Freud resorts to this behavior, when the stakes are considerably higher and I resist being bound by his pleasantry. I am thinking of the curious turn his professed (and no doubt real) embarrassment takes in *The Psychopathology of Everyday Life* as he winds up his explanation of three patent errors in the dream book. By these slips, we have seen, he betrayed not ignorance but his untruthfulness on other scores. But before he is done, he ventures to generalize about them. The errors occurred precisely when he was deliberately censoring the facts, holding back from his readers the complete analysis and consciously engaged in "distortion or concealment." This of course is quite a different argument from saying that a given error was motivated by a specific unconscious wish, because Freud is now adding that he must have had an unconscious wish to trip himself up. This wish has an element of predictability about it—a persistent wish. Freud treats it as a rule and modestly does not examine its motivation, which is flattering to him. "What I wanted to suppress often succeeded against my will in gaining access to what I had chosen to relate"—as opposed to what he chose to suppress—"and appeared in it in the form of an error that I failed to notice" (1901, 218–19). Then, after narrating a fourth error (not from the dream book), he produces the pleasantry I refer to:

> It may, in general, seem astonishing that the urge to tell the truth is so much stronger than is usually supposed. Perhaps, however, my being scarcely able to tell lies any more is a consequence of my occupation with psycho-analysis. As often as I try to distort something I succumb to an error or some other parapraxis that betrays my insincerity, as can be seen in this last example and in the previous ones. (1901, 221)

This pleasantry is undeniably witty, remarking as it does the irony that falsehoods speak true. Above all, it is a performance, which may indeed seem "astonishing" when one stops to think how many parts Freud is playing at once. He is the blunderer who committed those errors and the detective who has found himself out; the writer who consciously distorts his meaning yet is helplessly sincere; the willing liar who cannot help telling the truth. The hidden motive is now out—"the urge to tell the truth." But what a strange motive for the Freudian unconscious! Quite different from the usual revelations of psychoanalysis, which turn inside out in a kind of satire on the human condition, this revelation is like that of an entertainment for children in which a disreputable exterior conceals a heart that can be trusted. The short paragraph modulates from a general skepticism about truth telling to the pleasing discovery about this one individual who, try as he might, cannot deceive anyone. Freud redeems humanity from the general error; or, as he would have it, psychoanalysis has redeemed him.

Don Quixote was clearly one of Freud's treasured books (1960, 43–46). In one place Cervantes has his hero remark, "To be witty and write humorously requires great genius," and "The most cunning part in comedy is the clown's, for a man who wants to be taken for a simpleton must never be one" (1605–15, 550). The above performance is a form of clowning, I submit, because Freud exaggerates his whole being, as well as these particular errors, in order to pull it off. He jokingly presents himself as a habitual liar, who then helplessly undergoes correction. His very efforts to be otherwise typify the automatism of a clown, for "as often as I try . . . I succumb." Good clowning requires abilities well beyond the ordinary. But because Freud is not a clown by profession and is in the present case writing a book about certain manifestations of clumsiness in everyday life, one may ask what he gains by this humor. Though veiled as a joke, the pleasantry about truth telling is actually quite ambitious. The effect is to make acceptable the evidence that he witnesses to, both in his discussion of these errors and in his wider writings. Or in respect to his character, the purpose is to assure the reader of his sincerity once the foolery is over. But Freud's way of seeking approval cheerfully admits to telling lies and speaks of insincerity. His clowning accommodates all the possibilities. If you enjoy the act, you identify not only with the positive assurances but with the scurrilous admissions.

The matter may finally be too serious for this kind of act. The stakes are high, if the very evidence Freud adduces and his sincerity in presenting it are in question. Though he may protest that he is joking, he wants me to accede both to the positive purposes and to the lies and insincerity. The performance takes my breath away, not because I am delighted but because I am terrified that I am the only person in the audience who recalls that just moments before Freud has indicated that the lies in question were not unconscious but committed in the course of deliberate "distortions or concealment" during the writing of his previous book. I find the confession and its appeal fascinating, but I cannot quite laugh. To paraphrase Freud, perhaps my being scarcely able to laugh any longer is a consequence of my occupation with psychoanalysis.

Freud has many ingenious ways of disarming criticism. As Frank Cioffi has painstakingly shown (1970), the means of avoiding refutation in Freud's writings can amount to a methodology in themselves. Some of the rhetorical rearguard action of *The Interpretation of Dreams* I have already noted in passing, and of course one of my principal contentions is that the pleasures of the book, including wish fulfillment in the argument thereof, carry the author and his readers uncritically forward. Half of the dream work itself seems curiously constructed to forestall objections to the theory. As compared to condensation and displacement, in truth, the so-called considerations of representability and secondary revision are more like window dressing. That Freud himself was a little uncertain how many classifications of dream work were useful is evident from their very tentative emergence. According to the rubrics created in sections A, B, D, and I of chapter 6 and the opening sentence of the last (488), there are four sorts of dream work; but Freud mainly relies on displacement and condensation, and in a couple of places the classificatory scheme noticeably requires anchoring to an editorial footnote (343n, 445n). Considerations of representability would seem to be circular and thereby a perfectly safe idea about the dream work, in that whatever is represented must be in representable form. Under this rubric, Freud has acute things to say about the difference of words and pictures as well as the inherent metaphoricity of language. If "linguistic usages, superstitions and customs" (347) are to be consulted in interpreting dreams, however, the significance of social history for the history of psychoanalysis is all the more apparent. The circularity of the discussion becomes unmistakable when the "Representation of Symbols" section makes its appearance in later editions. If, for example, "there is no group of ideas that is incapable of representing sexual facts and wishes" (372), considerations of representability cannot be worth considering.

As for the much-disputed secondary revision, Freud himself seems to have decided that it never should have been included as dream work (1923, 241). Yet it is not hard to see why he entertained the idea of secondary revision in 1900 and retained it in successive editions of his book. Freud's approach to dream interpretation, from chapter 2 onward, is piecemeal. Even a coherent manifest dream is taken apart piece by piece until (in the classic manner of constructing a narrative from circumstantial evidence) the latent content with its hidden meaning becomes clear. Some dreams are so fragmentary, or recalled in such a fragmentary way, that constructing a coherent story is the only method of interpretation that presents itself. Fundamental to Freud's theory—and to the entire enterprise—is that all dreams are disguised. (The exceptions that he notices hardly signify for the rest of the book.) Yet some manifest dreams persist in exhibiting coherence, so this appearance—not being the true dream—must be a product of the dream work. Hence secondary revision comes into play. As if to counter the tendencies of displacement toward incoherence, this process shapes the dream "to the model of an intelligible experience," introduces a modicum of logic and reasonableness "akin to waking thought," and constructs stories with unsurprising conclusions. Interspersed with this definition are cautions not to be fooled by dreams subjected to secondary revision: such dreams "appear to have a meaning, but that meaning is as far removed as possible from their true significance." What Freud is worrying about, obviously, are dreams that do not require analysis, "dreams which might be said to have been already interpreted once, before being submitted to waking interpretation" (490). Such dreams might be called in evidence for an entirely different theory, according to which, just as in waking thought, order and completion are aims not always achieved but usually desired. A caveat, therefore, becomes necessary: "For the purposes of *our* interpretation it remains an essential rule invariably to leave out of account the ostensible continuity of a dream as being of suspect origin, and to follow the same path back to the material of the dream-thoughts, no matter whether the dream itself is clear or confused" (500). Not surprisingly, in additions to the book in 1914, Freud disparages secondary revision as "the one factor in the dream-work which has been observed by the majority of writers on the subject" and already hints that its importance has been "over-estimated" (501–2). Secondary revision was apparently introduced, at the end of the long chapter on the dream work, to forestall the criticism that some dreams do not need interpretation, since they are not made up very differently from the way waking experiences are made out. Freud is cautious and then halfhearted about his last-minute contribution because it opens out toward a theory that would adopt coherent dreams as its prime examples. As usual, he would rather advance and retreat than surrender the claim that his singular theory accounts for all dreams.

Secondary revision, tacked onto the end of the chapter on the dream work, has a last-minute quality not unrelated to the manner in which Freud raises the "practical" issues bearing on his theory only at the end of the book. The latter performance is more brilliant, more light and appealing, than the halfhearted attempt to patch together the dream work—though I have only been able to characterize it, and the presence of the two stories of the censorship, as a resort to clowning. All the ways Freud has of protecting himself are consistent with the worst scenario that Clark Glymour projects from his low assessment of *The Interpretation of Dreams*—namely, that the whole book was a kind of ruse, to cover Freud's failure to come up with any evidence for his psychological theories. "At the turn of the century Freud once and for all made his decision as to whether or not to think critically, rigorously, honestly, and publicly about the reliability of his methods. *The Interpretation of Dreams* was his answer to the public, and perhaps to himself" (1983, 70–71). But Glymour is wrong to say he is putting psychology aside when he makes this conjecture and wide of the mark when he accuses Freud of "a weakness of intellectual will." If there is any truth in his conjecture (and I would not be citing it if I did not believe that the dream book was in some sense the instauration of a myth), then Freud's strength of intellectual will has seldom been surpassed, for what else but *will* secured his masterpiece and his fame? Nonetheless there is a certain waywardness—a diffidence somewhat coy, a fineness touched with shame—that could suggest the author knew he would not be engaging in science after this. The main thing that Glymour fails to address is how the book could achieve such ascendancy on these terms.

The reputation of Freud right now is not a very comfortable one. He may be a giant—he is a giant—but the arguments on which his gianthood rests have increasingly been challenged, not over this or that matter but as if he could be toppled. In a series of hard-hitting essays and reviews, Frederick Crews has effectively marshaled the doubts of American scholars less outspoken—and less rhetorically gifted—than himself. Crews has sensed the radical implications of Glymour's distrust of the dream book: the masterpiece that achieved so much can be turned about-face. "Freud was already a pseudoscientist from the hour that he published *The Interpretation of Dreams*." With some exaggeration Crews writes, "Though Freud loved to sermonize about courageously opposing the human penchant for self-deception, it is no exaggeration to say that his psychoanalytic career was both launched and maintained by systematic mendacity." He then retreats far enough to admit that Freud was not "a congenital liar," but never retracts his conclusion that the course taken by Freud was deliberate: "Given [Freud's] determination to advance a doctrine that had nothing to be said in its scientific favor, he had no alternative to resorting to legend building" (1986, 101–5). The uncompromising tone of these charges will not take prisoners of the remaining faithful, but Crews has also vigorously supported the work of other serious critics of psychoanalysis. In the matter of Freud's possible lying, Allen Esterson (1993) has gone considerably farther. By insistent comparison of one text to another, this page to that, and by repeated examination of the relation of Freud's reasoning to the evidence he adduced, Esterson makes out a powerful, if largely circumstantial case that Freud could not resist resorting to fiction on occasion, nearly always exaggerated his findings, and either blinded himself to self-contradiction or carefully buried it in rhetoric.

If such charges are even partly true, they underline the need for coming to terms with Freud's extraordinary influence today. Crews calls for an assessment of "what, if anything, is salvageable" and what "could ever have commanded our blind assent" (1986, 86), but only at the close of one of his reviews. A book by another generalist, Ernest Gellner, addresses more directly the why and wherefore of psychoanalysis, though not very satisfactorily so, since it is written at too great remove from Freud's texts. Gellner appreciates that the claim to science was as important to Freud as his gift for interpretation. He has coined the term *bio-hermeneutics* as descriptive of the combined appeal of a not conspicuously logical set of ideas. "Bio-hermeneutics is simultaneously reductive (thereby giving control) and restorative (dignity-preserving)." Whereas "pure science would be too abstract and distant, pure hermeneutics would not have any mysteries and inspire no awe. Bio-hermeneutics contains just the right mixture of threat and familiarity" (1985, 109–11). The right mixture, I take it, because it worked. Gellner here applies to psychoanalysis an insight he expressed earlier in an essay on ideologies, which, in general, sustain belief by a combination of promise and threat—the latter an "offense-generating property" that amounts to a demand with menaces, in the language of robbery or blackmail statutes (1979). That Freud himself began to treat psychoanalysis as an ideology is evident from "On the History of the Psychoanalytic Movement," which concluded ringingly, "Men are strong so long as they represent a strong idea; they become powerless when they oppose it" (1914, 66). There is no such threatening note, however, in Freud's first great book, which is all promise. As I hope I have shown, his principal means of swaying belief in *The Interpretation of Dreams* are pleasure-yielding. The drama of detection—especially when secrets are construed as something closely defended and not just things of which we are unaware—and the inevitable routing of the censorship manifest power, no doubt, but do not entail coercion. A shared knowingness and identification with the author of the dream book, with his modest ambition and less than heinous faults, are amply pleasing; only an occasional testiness or, more subtly, the clowning may be offensive to some readers. So also is the whole purpose of establishing dream interpretation as a science positively promising—unless one judges that for other readers science itself is intimidating.

Adolf Grünbaum notoriously expends a third of his book demanding Freud back from hermeneutics and the rest of it demonstrating that Freudian science is not proven (1984). Grünbaum has received far greater hearing from psychoanalysts than any other critic from outside the institutes in recent years, mainly because he takes the issue of scientific validity so seriously but also perhaps because his exposition is so dense that the implications register only very slowly. David Sachs (1989) contends that Grünbaum closes too quickly on a given passage without considering well what Freud wrote elsewhere, sometimes only a few pages away. The phenomenon of the many-sidedness of Freud's arguments has been addressed by Cioffi (1970), however, and now by Esterson (1993) without much comfort for psychoanalysis. The phenomenon is another reason for concluding that

Freud's writing of the dream book was an act of interpretation rather than a scientific endeavor. I believe the real difficulty for Grünbaum is Freud's readiness—as in the stories of Careless and Naughty—not to confine himself strictly to facts. An exhaustive critique of Freud's claims to empiricism would seem to be seriously compromised if one cannot be sure which clinical data were reported and which were invented for the sake of the story—to say nothing of the theory. Unfortunately, this frustration must extend to the wisest of defenses against Grünbaum's critique as well, notably Marshall Edelson's cumulative reflections on the theory of psychoanalysis (1988). If Freud's evidence is not only selective but eked out with pretended fact, then neither attack nor defense can be undertaken enthusiastically on scientific grounds.

Meanwhile quite a few observations by first-generation psychoanalysts fill one with fresh misgiving today, as with these words of Ernest Jones about the founder:

> Freud hardly ever indulged in controversy: it was distasteful to his nature. Like Darwin, and unlike most men of science, he responded to criticism, sensitive as he was to it, simply by continuing his researches and producing more and more evidence. He had little desire to influence his fellow men. He offered them something of value, but without any wish to force it on them. He disliked debates or even public scientific discussions, the object of which he knew was mainly controversial, and it was in deference to this attitude that papers read at psychoanalytic congresses have never been followed by any discussion of them. (1957, 1.31)

The first sentence of this extract is refuted by Jones's biography itself. The second generates unease about what it meant to produce evidence and may prompt cynics to ask what happened to it. The third and fourth are perfect examples of the trope of modest ambition, here adopted by the subject's biographer. But the fifth, with its casually dropped information about the practice of psychoanalytic congresses—in deference to Freud's modesty—is most stunning to an outsider today. After the mental wrestling of Grünbaum and others over the validity of the Freudian arguments as science, this acknowledgment that papers were never discussed rings a sad bell of futility, and the audience may depart before the next round.

In my judgment, the bell that tolls the passing of psychoanalysis as science is Sulloway's big book (1979)—which was partly inspired by Ellenberger's (1970). By redirecting attention to medical and scientific activity at the turn of the century, Sulloway sought to correct the received idea that the movement originated from Freud's self-analysis. His true achievement has been—by careful research and patient detail—to historicize Freud's thinking more thoroughly than has ever been done in a single volume. Sulloway has not attacked; he has surrounded Freud, placed him in the midst of science that is palpably dated. Freud's friend Fliess, for example, comes across with more dignity than he is accorded by most Freudians, simply because Sulloway presents a context for Fliess's theories in the science of the time. So, too, he treats Freud, never disrespectfully but inevitably causing him to slip further into the past. I cannot agree with Paul Robinson (1993, 18–100) that Sulloway's is an elaborate, roundabout effort to sully Freud's reputation. Robinson four or five times charges his fellow historian with "latent" hostility and persistently intimates that *Freud, Biologist of the Mind* pretends to be something other than it is. This is a form of intellectual blackmail (I could reveal your secret motive) that Freud himself cautioned against, precisely because analysis can easily be misused so (1914, 49)—though it cannot be said that he always heeded his own advice in this regard. Besides, Robinson rather carelessly assumes that Sulloway wants to tar the founder of psychoanalysis by associating him not only with Fliess but with Darwin. Why would Sulloway—or Freud, for that matter—feel that association with Darwin was denigrating?

Patricia Kitcher has argued at length that *The Interpretation of Dreams*, consistent with the rest of psychoanalysis, is "an instance of interdisciplinary theory construction"—a procedure that she regards as scientifically appropriate in the circumstances. Kitcher specifically denies that Freud was a pseudoscientist and accepts in good faith his intention eventually to cover his bets and make good on the theory as a whole. Her critique of the dream book, in fact, is kinder to its theoretical chapter than to individual dream interpretations, which seem "badly underdetermined and poorly supported" (1992, 114, 143). It is no objection that Freud's book is theory-driven if the theories in question are promising; and since Freud could not expect neurophysiology to confirm his hypotheses, he "looked . . . to psychiatry, sexology, anthropology, and linguistics." To be sure, the resulting argument is "something of a mess."

> Given Freud's interdisciplinary approach, however, this messy appearance would not be disheartening. Using details from various mental or social sciences to fill in a neurophysiologically grounded theory skeleton would be perfectly reasonable, if there were good reason to believe that these sciences would all be part of a complete and integrated theory of the mind. If each of the pieces was right, then eventually they would all fit together. (1992, 120)

Thus Kitcher makes plenty of room for the dream book in the development of psychoanalysis. But if her critique is more eclectic than Sulloway's, it is also more conclusive, for she believes the interdisciplinary approach failed: Freud and psychoanalysis were unable to adapt to changes in biology and the social sciences. Kitcher tells this story, in part, as a caveat to present-day cognitive science.

Meanwhile Crews has taken Sulloway to task for being too kind, and the latter has returned for a second look at Freud's clinical papers, a little aghast at what he finds there (1991). The truth is that if Sulloway had not been largely sympathetic and felt drawn to Freud's ideas in the first place, he would not likely have devoted the kind of work to his book that the job demanded. What is needed now is an equally careful book, or several books, on the social history (much harder to document) immediately affecting psychoanalysis and feeding into it. One would suppose that the writings of Michel Foucault could partially fulfill this need, but though Foucault's own reasoning is unmistakably indebted to Freud, he does not seem to have wanted to place Freud's thinking itself in a nineteenth-century context. Though Foucault believes that the modern West is a confessional society, and sex "a privileged theme of confession," for example (1976, 58–68), he does not venture to say that Freud was a product of this society. A vast amount of material from the Victorian era—diaries and journalism, serious and popular treatises, high and low literature—has been gathered by Peter Gay (1984; 1986) with an eye to the contextualizing of psychoanalysis. Criminology and trial narratives might also be profitably studied, along with the literature based on them. We can very much use studies that trace the social history of *The Interpretation of Dreams* and later works in as great detail as possible.

Novels and plays are likely to provide the fullest documentation of this kind, since they register not only conditions of the time and rules of behavior but anxieties both public and private—and imagined happiness. A usual procedure has been to track down the books alluded to by Freud and reexamine them in light of the surrounding argument in the Freud text. Alexander Grinstein (1968), William J. McGrath (1986), and Peter Gay (1990, 95–124) have all had success with this approach. It would be still more valuable to step back from these titles and ask in what ways mainstream nineteenth-century literature actually shaped Freud's thinking. One ought to take the position that novels likely to have been encountered by Freud before 1900 contributed to psychoanalysis, and not merely that psychoanalysis turns out to be a good means of interpreting novels. Freud did not have any influence on Balzac and Dickens, George Eliot or Dostoevsky; it could only be the other way around. Ned Lukacher (1986, 330–36) has located in Freud's writing two motifs from *David Copperfield*, being born with a caul and biting Murdstone's hand; but in truth numerous actions of this favorite Dickens novel bear thinking about, not least the hero's self-presentation and modest ambition. Ronald R. Thomas (1990) has proposed still other harbingers. And of course not single novels but the conventions that they embrace matter most. Nineteenth-century novels repeatedly draw lines around what is proper and what is scandalous. Characters engage in damage control but usually, like the censorship, in vain. Attending to these wider concerns should help distinguish the bourgeois Freud from the romantic. If anything, too much attention has been paid to the rebellious cast of the unconscious in Freud's writings and not enough to the grounds of censorship, which are social. Freud did not play up rebellion, not even the rebellion of the sexual instincts; time after time, and especially in *The Interpretation of Dreams* and *The Psychopathology of Everyday Life*, he focused on distinct embarrassments. One does not have to deny his serious purposes to see the parallel to nineteenth-century playwrights and novelists, who also treated scandal entertainingly, even as they examined its serious consequences.

Once the social bearings of Freud's theorizing are better charted, his project may seem to relate to the nineteenth rather than the twentieth century, as it has already seemed to subside into a dated science because of Sulloway's work. For reasons having only remotely to do with psychoanalytic therapy, I believe that some additional perspective would be salutary. With too many humorless intellectuals, teachers, and others, the unconscious has literally become the place where anything goes—any argument, that is. The Freudian distinction between manifest content and latent content has been all too wishful a discovery if the true meaning of a text is always the latent meaning. If one is fond of this way of reasoning and patient enough, evidence can always mean what one wants it to mean. That which is repressed must signify more than that which is expressed—for as the character in Dickens was supposed to say, "What's his motive?" Fragments recovered from all that are forgotten weigh more heavily than what can readily be remembered. The slightest evidence is the most revealing. Nothing is accidental; everything is both a cause and an effect. The unconscious brooks no

negation or contradiction; logic is always imposed, counterpro-
ductive, and unnatural. Needless to say, almost any proposition
can be defended on principles such as these, and Freud may be
partly to blame for them because of his own eager manipulations
and incautious rulings, as in the following censure of doubt:

> Doubt whether a dream or certain of its details have been
> correctly reported is once more a derivative of the dream-
> censorship, of resistance to the penetration of the dream-
> thoughts into consciousness. This resistance has not been
> exhausted even by the displacements and substitutions it
> has brought about; it persists in the form of doubt attach-
> ing to the material which has been allowed through. We are
> especially inclined to misunderstand this doubt since it is
> careful never to attack the more intense elements of a
> dream but only the weak and indistinct ones. As we already
> know, however, a complete reversal of all psychical values
> takes place between the dream-thoughts and the dream. . . .
> That is why in analysing a dream I insist that the whole
> scale of estimates of certainty shall be abandoned and that
> the faintest possibility that something of this or that sort
> may have occurred in the dream shall be treated as com-
> plete certainty. (515–16)

The trouble is that many people who have become inured to
such antiskepticism are not dealing with dreams or thinking of
psychoanalysis but struggling with other intellectual tasks, for
which it may easily become a formula for bullying or self-deception. All such formulas should be scrutinized anew, and this will
occur as Freud's texts are forthrightly reviewed and placed in
historical perspective.

How did it come about that into the unconscious entered all
those thoughts judged wicked or improper at the time? And why
did it come about? What space needed filling? What things were
not repressed? Why were the petty fancies of personal ambition
stashed away when far more ambitious plans and actions, such as
writing a masterwork on dreams, were not repressed? Why were

thoughts rather than actions the substance of Freud's life work? History should hold the answers to these questions, which I believe were on Carl Schorske's mind in the somewhat cryptic close to his essay on the politics of *The Interpretation of Dreams*: "By reducing his own political past and present to an epiphenomenal status in relation to the primal conflict between father and son, Freud gave his fellow liberals an a-historical theory of man and society that could make bearable a political world spun out of orbit and beyond control" (1973, 203). Notoriously, some of the manipulations of manifest and latent meaning that I have listed above, which aim to surprise audiences with their counter-intuitive results, are characteristic of "theory" in a variety of disciplines today, when many feel politically powerless again and scholars struggle to keep up with a bewildering knowledge explosion. Theory sometimes takes refuge in knowingness, akin to the knowingness that is a source of so much pleasure in the dream book.

If we examine the habits that Freudian argumentation appears to sanction, Freud's reputation may be the better for it over the long run. *Die Traumdeutung* was hardly humorless theory building; even its author's self-importance was expressed as something of a joke. History, moreover, is not likely to forget his master-piece. In two substantial areas the positive achievement of the book seems indisputable: it endorsed the principle—especially apparent in its autobiographical passages—that each individual is constrained by his or her personal history, and it worked out an especially influential account of narrative. In neither the developmental thesis nor the analysis of narrative had Freud invented something entirely new, but in both areas he focused the attention of the twentieth century so decisively that they well deserve to be known as Freudian innovations. On the grounds of publicity, obviously, the unconscious also deserves to be called a Freudian innovation—for which a context has been filled in by Ellenberger (1970). I still have deep reservations about celebrating the prominence given by Freud to unconscious motives, a prominence that may sometimes tempt us these days to favor

strained inference over direct observation. Thus an object missing from a painting, for example, may be thought to loom more
significantly for interpretation than the objects depicted there,
or liberal political programs may be censured, because thought
to disguise other ends, more than undisguised aggression against
others; whereas I believe most of the civilized world's problems
are all too evident but prone to be ignored because of conflict-
ing, conscious motives.

Freud cannot be held responsible for the habit—insofar as it
is a habit—of seeking unconscious motives, but undoubtedly his
rhetoric exaggerated the force of unconscious ideas. As Paul
Robinson—the historian who attacks Sulloway—demurs, "All
Freud needed to prove was that *some* dreams and slips can be
explained by the assumption of unconscious motives" (1993,
232). But of course Freud did not content himself with staking
out less than universalist positions in the dream book, and this
practice handsomely paid off. Like most of us, he was more
tempted by irrationality—his own or that of others—than
moved by his unconscious. When, for example, in *The Interpretation of Dreams* he anticipates that his assertion that "*every*
dream is the fulfilment of a wish" will face "the most categorical
contradiction" but nonetheless contends that "there is no great
difficulty in meeting these apparently conclusive objections"
(134–35), he is carried away by daring, not some unconscious motive. As Robinson would agree, Freud knows very well that the
objections are rational, for he has raised them himself, but he is
determined to override opposition. Why? It is exhilarating to
make sweeping claims, and the immediate payoff may be credit
for a universal theory of dreams, the discovery of a scientific law.
Just so, we often realize—individually or collectively—that certain choices, policies, and struggles are ineffective or even coun-
terproductive, yet while engaged in argument we cannot bring
ourselves to change. The motives for such mental intransigence
may range from ambition to inertia, but introspection usually
reveals the contradictions to be quite conscious.

Developmental concerns in psychology would seem to branch off from Darwinian and pre-Darwinian evolutionism, as adapted by Ernst Haeckel and Freud himself. The stress on an individual's history has roots as well in romantic biography and nineteenth-century pathology. The practice of medicine can hardly avoid noticing that afflictions strike some individuals but not others. The cause of a disease may be generalizable, but who has been affected is not: hence the doctor should establish a history, which includes the family, for each new patient. Freud appreciated this difference between medicine and most sciences, as well as the awkward truth of natural history that lies behind it: each organism is different, even though it may respond to given stimuli in predictable ways. Differences culminate over time; events leave their impression on the individual. In the dream book the chief evidence of this concerns Freud himself: because his father was much older than his mother, because he was uncle to a nephew a year older than himself, certain relationships arose and were construed in such a way. Certain events occurred to or around the child and remained in his conscious or unconscious memory thereafter—his, and not another's. The unique history of each individual has become a mainstay of psychoanalysis and also, because of Freud's writings, a twentieth-century commonplace. Confusion often arises on this account when it is asked whether psychoanalysis is a science. For example, some of the data gathered by Seymour Fisher and Roger P. Greenberg (1985), data on the reliability of Freud's theories, disappoint the authors, who seek testable conclusions derived from repeatable experiments. But unique case histories, the construction of which may be beneficial to the subject, do not provide such data.

Wittgenstein, we have seen, thought Freud provided "the sort of explanation we are inclined to accept," or an explanation with "the attraction which mythological explanations have" (1966, 43). In his subsequent conversation he exemplified a myth by the sentence "It is all the outcome of something that happened long ago" (1966, 51). What Wittgenstein features in this comparison of psychoanalytic to mythological explanations is not necessarily communal life but the temporal aspect of each individual experience. A certain segment of the person's recoverable history completes a satisfying narrative rather than establishes the cause of symptoms. Since psychoanalysis per se does not offer to administer drugs or to operate on the heart or brain, his analogy to the uncertain power of myths may be more vital than that to the uncertain power of medicine—that is, when the myth is truly individualized, for if it is immediately generalized, as in an Oedipus complex, the satisfaction becomes more like that of a religion. The satisfying explanations are not likely to be scientific, for science does not aim at individual solutions. Sebastiano Timpanaro (1974, 85) cites a point made by Ernst Cassirer about the difference of science from myth in respect to their powers of explanation. Science—Cassirer was thinking of the physical sciences—is more willing to accommodate accident in its world. For myth, on the other hand, "there can be no accidental event in the course of human existence . . . Everything has to be strictly determined." His observations apply rather well to Freud. In myth "we find, in a certain sense, instead of an absence of all causal explanation, a hypertrophy, a superabundance of the causal impulse, and . . . causality is based on intentions and acts of will as contrasted with rules and laws" (1956, 99). As I have suggested earlier, there is also a modern precedent for explanations based on intentions in the administration of criminal justice. And I remain convinced that without the information revolution of the nineteenth century and the concomitant enhancement of secrets, psychoanalysis would scarcely have been thought of by anyone (1985, 337–77).

Explanations of most sorts take a narrative form, and Freud not only expertly deployed narrative in *The Interpretation of Dreams* but provided a methodology for analyzing it. The book's purview is very nearly limited to narrative, as I have stressed by noting Wittgenstein's careful insistence on *hallucinated* wish fulfillments: Freud was writing not of dreams that come true but of dreams that minimally picture and typically narrate some degree of satisfaction. Two important steps for understanding narrative he took as one: he recognized the value of make-believe as evidence together with the most common reason for indulging in it—wish fulfillment. (I love Freud for that, for my own profession is the study of literary fictions and the realities from which they differ.) Again, the theory was not new. Since poetic license was one of the earliest human discoveries, it was not long before the Greeks began to theorize about it, and Francis Bacon can be taken as a modern instance of the tradition: in distinguishing imaginative from historical writing, Bacon wrote that poesy "doth raise and erect the mind, by submitting the show of things to the desires of the mind; whereas reason doth buckle and bow the mind into the nature of things" (1605, 83). Bacon's account, in short, addresses the difference between a pleasure principle and a reality principle. Yet Freud's masterpiece, more than any other work by his own or another's hand, drew attention to the principle of wish fulfillment and acquainted us with the narrative devices of desire. Though a hermeneutics of suspicion has many compelling precedents in modern times, Freud's wry skepticism encouraged us to suspect desire even in narratives supposed to be less subject to it than poesy is. (I respect Freud for this, because without his explorations I doubt if it would have occurred to me to see wish fulfillment as a guide to the argument of his dream book.)

Students of narrative often pass by the importance Freud ceded to make-believe and wish fulfillment in order to come to terms directly with the chapter on the dream work. What Freud called the dream work (Traumarbeit) is a set of operations nicely suited for transforming one or more narratives into a new one, and mapping such narratives one upon another becomes a powerful method of analysis. Displacement and condensation are transformations that can be studied profitably without necessarily tangling with loaded questions about how verbal representations are possible in the first place. If an analyst of narrative—and of course the analyst does not have to be a literary critic—can discern a basis for suspecting condensation or displacement, such as parallel actions in a plot, then the differences are revealing as well, because the several narratives have been invented by the same person. It is easiest to illustrate this with reference to the kind of displacement that Freud calls identification (Identifizierung), a term he mainly reserves for the various actors in a dream: "Thus my ego may be represented in a dream several times over, now directly and now through identification with extraneous persons" (323). Similarly, if in a novel a character named Steerforth seduces a working-class girl, and another named Heep lusts after an angelic woman associated with church windows and is outrageously ambitious to boot, we begin to surmise that these characters are displaced versions of the hero, who is unconsciously in love with both women and modestly hopes to rise in the world. Without analyzing the several story lines in this way, we would scarcely guess that Copperfield is a sexual being. Uriah Heep, because he has no viable role in the novel except as relating to the hero, is a true double or dreamlike identification. Thus a process comparable to the dream work explains the construction of this novel. Psychoanalysis, not surprisingly, copes beautifully with *David Copperfield*; conversely, the identifications in question are a function of specific constraints upon heroes and heroines in the nineteenth century without which psychoanalysis would not have been invented.

The operations grouped by Freud under considerations of representability and secondary revision equally apply to narrative, though not very effectively in the analysis of narrative, since it is not easy to reach agreement about symbolic representations or the dream logic. By association, for example, one thing may stand for its opposite; but such is not *always* the case—no more than contradiction is wholly inconceivable in dreams or analogous fictions. Secondary revision may be more aptly ascribed to literary than to dream constructions, since it would seem to resemble the shaping of deliberate artistry. Some psychoanalysts now believe that therapy itself operates on a plane of narrative independent of case study as it was formerly understood. Most nakedly, Donald P. Spence (1982) has urged that the construction of a story satisfying to the patient is as effective as ascertaining the sources of disturbing symptoms. Spence's view is congruent with my sense that psychoanalysis mainly describes narratives, but I doubt that a narrative can be satisfying if the patient does not believe it results in his or her true history. The problem is like that of psychoanalysis as a movement: it is not a science, but without the claim to be science it must lose interest for many people.

That brings me back to the jocose aspect of Freud's self-presentation, to the dream book as performance. I cannot believe that *The Interpretation of Dreams*, even its most egregious use of fiction, was meant to deceive in a humorless way or to garner the admiration of any audience but the willing. Its persistent knowingness, charged with the possession and revelation of secrets, is far less dreary than the knowingness it has spawned in much recent theorizing in my own and other fields. Freud's knowingness verges now on winking, now on the deadpan of clowns; or he emits the humor of great nineteenth-century novelists in their most serious project, for in general modern Europeans of that era preferred to read fiction that was presented to them in the guise of realism. Just as the novel therefore posed as anti-romance, a fiction-despiser's fiction, the dream book scorned dreams in favor of their true meanings and the technique of uncovering them. Such parallel institutions as the novel and psychoanalysis deserve more attention from historians.

Of course, Freud frequently credited the poets and novelists who came before him. As Mark Edmundson has argued (1990), he even usurped their authority in the act of his own self-definition. But those novelists and poets never relentlessly imputed unconscious motives to their characters at the expense of their studies of irrational behavior and the conscious sensations of public and private life. In truth the dream book, also, was not as exclusively concerned with the unconscious as Freud claimed, for most of the motives uncovered by analysis of his own dreams, beginning with the specimen dream itself, were such as likely to be regarded improper, or embarrassing if made public, rather than strictly unconscious. The novelists of the nineteenth century occasionally touched on the unconscious but continually dwelt on social proprieties and embarrassment. Because they thought of themselves as realists, they sometimes also disclaimed the ethical purport of their fictions, as Freud acquitted dreams at the end of his book. And many of them occupied themselves with what may be called the Oedipus complex.

Freud's contribution was not to explain something that, as he indicated in *The Interpretation of Dreams*, was already known to Sophocles and Shakespeare. Rather, he consulted his own experience of growing up in the nineteenth century and gave it a name that, as much as anything, has established his own fame in our time. In the dream book, his reflections on the death of his father relate closely to his analyses of the laboratory dreams—stories of personal and professional ambition—and turn upon thoughts that were far from unconscious. A modern industrial and market economy relies implicitly on the ambition of young people; so do families rely on that ambition. But families are also hierarchical in their distribution of power and respect, and were even more so in Freud's time. So if we wonder why Freud should have discovered a timeless and universal Oedipus complex just at the end of the nineteenth century, we may suppose that explaining competitiveness as infantile in origin assuaged his own feelings and at the same time provided a myth acquitting all those bent on surpassing their fathers. Infantile sexuality, too, is a comforting thought if adult sexuality is so discomforting that it requires explanation, but sexuality is not the most important theme, wishful or otherwise, of the dream book. Ambition is the principal theme, and Freud's masterpiece succeeds in interpreting ambition as wishfulness.

Anzieu, Didier. 1975. *Freud's Self-Analysis.* Trans. Peter Graham. London: Hogarth, 1986. (25)

Appignanesi, Lisa, and John Forrester. 1992. *Freud's Women.* New York: Basic. (25)

Bacon, Francis. 1605. *The Advancement of Learning.* London: Everyman, 1915. (135)

Bloom, Harold. 1989. *Ruin the Sacred Truths: Poetry and Belief from the Bible to the Present.* Cambridge: Harvard University Press. (13)

Cassirer, Ernst. 1956. *Determinism and Indeterminism in Modern Physics.* New Haven: Yale University Press. (134)

Cervantes Saavedra, Miguel de. 1605–15. *Don Quixote of La Mancha.* Trans. Walter Starkie. New York: Signet, 1964. (62, 118)

Cioffi, Frank. 1970. "Freud and the Idea of a Pseudo-Science." In *Explanation in the Behavioural Sciences,* ed. Robert Borger and Frank Cioffi. Cambridge: Cambridge University Press. (119, 124)

Collins, Wilkie. 1868. *The Moonstone.* Ed. J.I.M. Stewart. Harmondsworth: Penguin, 1966. (39–40)

Cranefield, Paul F. 1958. "Josef Breuer's Evaluation of His Contribution to Psychoanalysis." *International Journal of Psycho-Analysis* 39:319–22. (49)

Crews, Frederick. 1986. *Skeptical Engagements.* New York: Oxford University Press. (122–23)

Darwin, Charles. 1859. *On the Origin of Species.* Facsimile ed. Cambridge: Harvard University Press, 1966. (13–14, 30, 47–48)

Deleuze, Gilles, and Felix Guattari. 1972. *Anti-Oedipus: Capitalism and Schizophrenia.* Trans. Robert Hurley, Mark Seem, and Helen R. Lane. Minneapolis: University of Minnesota Press, 1983. (72)

Derrida, Jacques. 1966. "Freud and the Scene of Writing." In *Writing and Difference,* trans. Alan Bass. Chicago: University of Chicago Press, 1978. (17)

Edelson, Marshall. 1988. *Psychoanalysis: A Theory in Crisis.* Chicago: University of Chicago Press. (125)

Edmundson, Mark. 1990. *Towards Reading Freud: Self-Creation in Milton, Wordsworth, Emerson and Sigmund Freud.* Princeton: Princeton University Press. (137)

Ellenberger, Henri F. 1970. *The Discovery of the Unconscious: The History and Evolution of Dynamic Psychiatry.* New York: Basic. (37, 126, 131)

Erikson, Erik Homburger. 1954. "The Dream Specimen of Psychoanalysis." *Journal of the American Psychoanalytical Association* 2:5–56. (22)

———. 1968. *Identity: Youth and Crisis.* New York: Norton. (63)

Esterson, Allen. 1993. *Seductive Mirage: An Exploration of the Work of Sigmund Freud.* Chicago: Open Court. (123, 124)

Fisher, Seymour, and Roger P. Greenberg. 1985. *The Scientific Credibility of Freud's Theories and Therapy.* New York: Columbia University Press. (26–27, 88, 133)

Forster, John. 1872–74. *The Life of Charles Dickens.* 2 vols. London: Everyman, 1927. (40)

Foucault, Michel. 1976. *The History of Sexuality.* Vol. 1. Trans. Robert Hurley. New York: Vintage, 1980. (128)

Freud, Sigmund, and Josef Breuer. 1895. *Studies on Hysteria.* Standard ed. Vol. 2. (42, 79, 103–4, 105)

Freud, Sigmund. 1896. "Further Remarks on the Neuro-Psychoses of Defence." Standard ed. Vol. 3. (79, 98)

———. 1900. *The Interpretation of Dreams.* Standard ed. Vols. 4–5. (Passim)

———. 1901. *The Psychopathology of Everyday Life.* Standard ed. Vol. 6. (30, 53–54, 102–3, 117–18, 129)

———. 1905a. *Jokes and Their Relation to the Unconscious.* Standard ed. Vol. 8. (30, 91–93, 94, 99)

———. 1905b. "Fragment of an Analysis of a Case of Hysteria." Standard ed. Vol. 7. (36–37)

———. 1906. "Psycho-Analysis and the Establishment of the Facts in Legal Proceedings." Standard ed. Vol. 9. (36–37)

————. 1910. "A Special Type of Choice of Object Made by Men." Standard ed. Vol. 11. (74)

————. 1913. *Totem and Taboo*. Standard ed. Vol. 13. (68)

————. 1914. "On the History of the Psycho-Analytic Movement." Standard ed. Vol. 14. (31, 48, 124, 126)

————. 1915. "The Unconscious." Standard ed. Vol. 14. (42–43, 86, 98)

————. 1916–17. *Introductory Lectures on Psycho-Analysis*. Standard ed. Vols. 15–16. (9, 13, 38, 45–46)

————. 1917. "A Difficulty in the Path of Psycho-Analysis." Standard ed. Vol. 17. (13)

————. 1920. *Beyond the Pleasure Principle*. Standard ed. Vol. 18. (9)

————. 1923. "Two Encyclopaedia Articles." Standard ed. Vol. 18. (120)

————. 1925. "Some Additional Notes on Dream-Interpretation as a Whole." Standard ed. Vol. 19. (113–14)

————. 1927. *The Future of an Illusion*. Standard ed. Vol. 21. (29–30)

————. 1930. *Civilization and Its Discontents*. Standard ed. Vol. 21. (102)

————. 1933. *New Introductory Lectures on Psycho-Analysis*. Standard ed. Vol. 22. (85, 105–6)

————. 1936. "A Disturbance of Memory on the Acropolis." Standard ed. Vol. 22. (76)

————. 1940. *An Outline of Psycho-Analysis*. Standard ed. Vol. 23. (87)

————. 1940–52. *Gesammelte Werke*. Ed. Anna Freud, Marie Bonaparte, et al. 18 vols. London: Imago. (xii, 93)

————. 1950. *The Origins of Psycho-Analysis*. Standard ed. Vol. 1. (5, 22, 105)

————. 1953–74. *The Standard Edition of the Complete Psychological Works of Sigmund Freud*. Ed. and trans. James Strachey et al. 24 vols. London: Hogarth. (xii, 4n, 21, 25, 85, 93)

————. 1960. *Letters of Sigmund Freud*. Ed. Ernst L. Freud. New York: Dover, 1992. (56–57, 118)

————. 1965. *A Psycho-Analytic Dialogue: The Letters of Sigmund Freud and Karl Abraham, 1907–1926*. Ed. Hilda C. Abraham and

Ernst L. Freud. Trans. Bernard March and Hilda C. Abraham. New York: Basic. (25)

Freud, Sigmund. 1985. *The Complete Letters of Sigmund Freud to Wilhelm Fliess 1887–1904*. Ed. and trans. Jeffrey Moussaieff Masson. Cambridge: Harvard University Press. (20–21, 23, 27, 31, 33, 57, 102)

Gay, Peter. 1984. *The Bourgeois Experience: Victoria to Freud.* Vol. 1: *Education of the Senses.* New York: Oxford University Press. (128)

———. 1986. *The Bourgeois Experience: Victoria to Freud.* Vol. 2: *The Tender Passion.* New York: Oxford University Press. (128)

———. 1988. *Freud: A Life for Our Time.* New York: Norton. (3, 56, 69)

———. 1990. "Reading Freud through Freud's Reading." In *Reading Freud: Explorations and Entertainments.* New Haven: Yale University Press. (128)

Gellner, Ernest. 1979. "Notes Towards a Theory of Ideology." In *Spectacles and Predicaments: Essays in Social Theory.* Cambridge: Cambridge University Press. (123–24)

———. 1985. *The Psychoanalytic Movement: Or the Coming of Unreason.* London: Paladin. (123)

Gilman, Sander L. 1985. *Difference and Pathology: Stereotypes of Sexuality, Race, and Madness.* Ithaca: Cornell University Press. (94)

Glymour, Clark. 1983. "The Theory of Your Dreams." In *Physics, Philosophy and Psychoanalysis,* ed. R. S. Cohen and L. Laudan. Boston: Reidel. (15, 122)

Grinstein, Alexander. 1968. *On Sigmund Freud's Dreams.* Detroit: Wayne State University Press. (25, 69, 128)

Gross, Hans. 1905. *Criminal Psychology: A Manual for Judges, Practitioners, and Students.* 4th ed. Trans. Horace M. Kallen. Boston: Little, Brown, 1911. (37)

Grosskurth, Phyllis. 1991. *The Secret Ring: Freud's Inner Circle and the Politics of Psychoanalysis.* London: Cape. (48)

Grünbaum, Adolph. 1984. *The Foundations of Psychoanalysis: A Philosophical Critique.* Berkeley: University of California Press. (15, 22, 124–25)

Hacking, Ian. 1975. *The Emergence of Probability: A Philosophical Study of Early Ideas about Probability, Induction and Statistical Inference.* Cambridge: Cambridge University Press. (45)

———. 1990. *The Taming of Chance.* Cambridge: Cambridge University Press. (45)

Hobson, J. Allan. 1988. *The Dreaming Brain.* New York: Basic. (88)

Jones, Ernest. 1953. *The Life and Work of Sigmund Freud.* 3 vols. New York: Basic. (3, 39, 40, 48–49, 76, 125–26)

Kitcher, Patricia. 1992. *Freud's Dream: A Complete Interdisciplinary Science of Mind.* Cambridge: MIT Press. (126–27)

Krüll, Marianne. 1979. *Freud and His Father.* Trans. Arnold J. Pomerans. New York: Norton, 1986. (75–76)

Lacan, Jacques. 1956. *Speech and Language in Psychoanalysis.* Trans. Anthony Wilden. Baltimore: Johns Hopkins University Press, 1968. (17)

Lukacher, Ned. 1986. *Primal Scenes: Literature, Philosophy, Psychoanalysis.* Ithaca: Cornell University Press. (128)

McGrath, William J. 1986. *Freud's Discovery of Psychoanalysis: The Politics of Hysteria.* Ithaca: Cornell University Press. (22, 66, 68–69, 128)

Mandeville, Bernard. 1714. *The Fable of the Bees; or, Private Vices, Publick Benefits.* Ed. F. B. Kaye. Oxford: Clarendon, 1924. (104)

Mill, John Stuart. 1859. *On Liberty.* In *Collected Works of John Stuart Mill,* ed. J. M. Robson, vol. 18. Toronto: University of Toronto Press, 1977. (98)

Rand, Nicholas, and Maria Torok. 1993. "Questions to Freudian Psychoanalysis: Dream Interpretation, Reality, Fantasy." *Critical Inquiry* 19:567–94. (19)

Ricoeur, Paul. 1970. *Freud and Philosophy: An Essay on Interpretation.* New Haven: Yale University Press. (87, 88)

Riesman, David. 1950. "The Themes of Work and Play in the Structure of Freud's Thought." *Psychiatry* 13:1–16. (50)

Ritvo, Lucille B. 1990. *Darwin's Influence on Freud.* New Haven: Yale University Press. (14)

Robinson, Paul. 1993. *Freud and His Critics.* Berkeley: University of California Press. (126, 132)

Sachs, David. 1989. "In Fairness to Freud: A Critical Notice of *The Foundations of Psychoanalysis*, by Adolf Grünbaum." *Philosophical Review* 98:349–78. (124)

Schafer, Roy. 1977. "Problems in Freud's Psychology of Women." In *Female Psychology: Contemporary Psychoanalytic Views*, ed. Harold P. Blum. New York: International Universities Press. (106)

Schopenhauer, Arthur. 1844. *The World as Will and Representation*. Trans. E.F.J. Payne. 2 vols. New York: Dover, 1969. (42)

Schorske, Carl. 1973. "Politics and Patricide in Freud's *Interpretation of Dreams*." In *Fin-de-Siècle Vienna: Politics and Culture*. New York: Knopf, 1980. (22, 62–63, 82, 131)

Schur, Max. 1966. "Some Additional 'Day Residues' of 'The Specimen Dream of Psychoanalysis.'" In *Psychoanalysis—A General Psychology: Essays in Honor of Heinz Hartman*, ed. Rudolph M. Lowenstein, Lottie M. Newman, and Albert J. Solnit. New York: International Universities Press. (22–25)

———. 1972. *Freud: Living and Dying*. New York: International Universities Press. (24)

Spence, Donald P. 1982. *Narrative Truth and Historical Truth: Meaning and Interpretation in Psychoanalysis*. New York: Norton. (136–37)

Spock, Benjamin. 1946. *Baby and Child Care*. New York: Pocket Books, 1957. (59)

Sulloway, Frank J. 1979. *Freud, Biologist of the Mind: Beyond the Psychoanalytic Legend*. New York: Basic. (3, 14, 23, 126)

———. 1991. "Reassessing Freud's Case Histories: The Social Construction of Psychoanalysis." *Isis* 18:245–75. (127)

Thomas, Ronald R. 1990. *Dreams of Authority: Freud and the Fictions of the Unconscious*. Ithaca: Cornell University Press. (129)

Timpanaro, Sebastiano. 1974. *The Freudian Slip: Psychoanalysis and Textual Criticism*. Trans. Kate Soper. London: NBL, 1976. (134)

Welsh, Alexander. 1985. *George Eliot and Blackmail*. Cambridge: Harvard University Press. (87, 134)

———. 1987. *From Copyright to Copperfield: The Identity of Dickens*. Cambridge: Harvard University Press. (75)

————. 1992a *Strong Representations: Narrative and Circumstantial Evidence in England.* Baltimore: Johns Hopkins University Press. (37–38, 45, 77)

————. 1992b. "Patriarchy, Contract, and Repression in Scott's Novels." In *The Hero of the Waverley Novels.* Princeton: Princeton University Press. (77)

Winnicott, D. W. 1971. *Playing and Reality.* New York: Basic. (73–74)

Wittgenstein, L. 1966. *Lectures and Conversations on Aesthetics, Psychology and Religious Belief.* Ed. Cyril Barrett. Berkeley: University of California Press. (10–13, 14–16, 34–35, 44, 133–34)

Wollheim, Richard. 1971. *Sigmund Freud.* Reissued with a supplementary preface. Cambridge: Cambridge University Press, 1990. (3)

————. 1990. Supplementary preface to *Sigmund Freud.* Cambridge: Cambridge University Press. (41–42)

Yeazell, Ruth Bernard. 1991. *Fictions of Modesty: Women and Courtship in the English Novel.* Chicago: University of Chicago Press. (57)

Alexander Welsh is Emily Sanford Professor of English at Yale University. Among his books are *Reflections on the Hero as Quixote* (Princeton), *The Hero of the Waverley Novels* (Princeton), *George Eliot and Blackmail* (Harvard), and *Strong Representations: Narrative and Circumstantial Evidence in England* (Johns Hopkins).